Pig Environment Problems

PIG ENVIRONMENT PROBLEMS

P Smith

Pork Chain Solutions Ltd and PIGSPEC

H Crabtree

Farmex Ltd

NOTTINGHAM
University Press

Nottingham University Press
Manor Farm, Main Street, Thrumpton
Nottingham, NG11 0AX, United Kingdom

NOTTINGHAM

First published 2005
© P. Smith and H. Crabtree

All rights reserved. No part of this publication
may be reproduced in any material form
(including photocopying or storing in any
medium by electronic means and whether or not
transiently or incidentally to some other use of
this publication) without the written permission
of the copyright holder except in accordance with
the provisions of the Copyright, Designs and
Patents Act 1988. Applications for the copyright
holder's written permission to reproduce any part
of this publication should be addressed to the publishers.

British Library Cataloguing in Publication Data
Pig Environment Problems

ISBN   1-897676-18-2

Disclaimer

Every reasonable effort has been made to ensure that the material in this book is true, correct, complete and appropriate at the time of writing. Nevertheless the publishers, the editors and the authors do not accept responsibility for any omission or error, or for any injury, damage, loss or financial consequences arising from the use of the book.

Typeset by Nottingham University Press, Nottingham
Printed and bound by Hobbs the Printers, Hampshire, England

CONTENTS

1	Pig environment problems	1
2	Social factors	43
3	The importance of thermal comfort, light and electrical safety	71
4	The impact of noxious gases	85
5	Dust - a problem for pigs and people	95
6	Practical considerations	109
7	Using air movement to improve pig environment	125
	References	157
	Index	161

Dedication

To those who care for pigs and strive to provide consumers with wholesome pork.

Acknowledgements

The authors are pleased to acknowledge helpful input from pig farmers and a wide range of individuals and organisations involved in the diverse global pig industry. In particular, special thanks are due to Nick Bird of Farmex and Peter Maas of InterContinental B.V. for advice on ventilation issues. Information on pig health and related issues was generously provided by Elanco, Pfizer Animal Health and the Responsible Use of Medicines in Agriculture Alliance.

It was only possible to write Pig Environment Problems having been generously furnished with both published and unpublished information from many pig researchers throughout the world. We have particularly relied on work undertaken by Professor Chris Wathes and colleagues at Silsoe Research Institute. Special thanks are also offered to Jamie Robertson of the Centre for Rural Building, Aberdeen for steering us through a range of research on air quality particularly the findings of The International Commission of Agricultural Engineering.

The authors wish to acknowledge the use of reports and data supplied by the Meat and Livestock Commission's British Pig Executive and also from Danske Slagterier the Danish Bacon and Meat Council. and the Dutch Meat Board.

Thanks are also due to the Directors of Pork Chain Solutions Ltd for helpful comments and also to Beverley Goldthorpe for her meticulous typing. We would like to thank many colleagues in the pig industry for generously supplying photographs, not all of which we were able to use in the book.

Finally, we would like to express our thanks to our publishers for their patience and understanding when other pressures have taken us away from book writing, and to our families for tolerating our inexplicable obsession with the comfort of pigs.

1

PIG ENVIRONMENT PROBLEMS

Introduction

Pigs and people both thrive in good environments and struggle when their environments are inadequate. Within the United Kingdom, during a typical working day, around ten thousand farm workers share their air space with pigs for long periods. At the end of the working day the labour force discards protective clothing, scrubs down and takes refuge in a pig-free zone and relishes the opportunity to breathe clean air. Most pigs remain within a confined air space 24 hours per day. The first hint of daylight and whiff of fresh air, for many pigs, is restricted to that brief interval between pig housing and haulier's lorry.

Both people and nature exert impossible demands on the climatic environment of the pig. Whereas photographs taken on the right day promote a perception of an idyllic life for the outdoor pig, the reality can be very different. In its natural environment the pig has to cope with seasonal temperature fluctuations which mimic the extremes of an ice-capped mountain and the scorching heat of the desert. On some days pigs are subjected to such extremes within a matter of hours. Sometimes, within days, pigs have to cope with monsoon like rainfall and the dusty clouds associated with drought-like conditions. Physiological and behavioural flexibility are key requirements of productive pigs. These attributes have to be matched by demanding resilience in those charged with caring for pigs.

Pigs are expected to thrive under humid leaden skies and remain physically productive and impeccably behaved when lying on a wet floor and confronted by a chilling wind. During a twenty week time span, characterised by rapid growth, further demands are created since pig bodyweight increases up to seventy-fold and the associated heat output up to twenty-fold.

Pig keepers have very successfully attempted to overcome such challenges by bringing pigs indoors, confining them within well-defined areas, providing bedding and insulation in draught-free conditions and providing facilities for increasing air speed at pig level in hot weather. In effect, pigs demand the impossible and good pig keepers strive to provide it for them, indoors and outdoors.

Farm animals are primarily produced and reared for human food provision. In an age of plenty, the market place abounds with choice. Consumers demand a wide variety of food to suit all tastes and needs. Many consumers reasonably expect that their food should be totally traceable, having been produced in safe, clean, contamination-free environments which are routinely monitored and finely-tuned as required. Furthermore, there are additional concerns about the welfare of meat-producing animals and these are fast becoming global issues.

Consumer demands for low-cost pig meat have forced the industrialisation of pig keeping. In the interests of labour and production efficiency, pigs have been crowded into smaller air spaces where noxious gases wreak havoc with pig environments. Furthermore, many farming people, who yearned for an outdoor life, specialised by demand and found themselves working indoors to provide consumers with pig meat at affordable prices. Consumers are fickle. Faced with an array of choice, their demands change. The market place now tells pig keepers that increasing numbers of consumers are not comfortable with their perception of intensive pig keeping. Surveys consistently indicate that many enlightened customers declare they would be prepared to pay a premium for pig meat produced in softer environments. The actual buying habits of many consumers fail to confirm this stated intention. These changes in consumer demands coincide with extra demands on the pig industry under the Health and Safety at Work Act. Furthermore, given the diminishing labour force within farming, it seems likely that if quality labour is to be recruited, trained and retained, people working with pigs will demand better working conditions. There are therefore good reasons for increasingly focussing on pig environment.

"Wholesomeness" has become a buzz-word in meat production. The widespread use of antibiotics, particularly in intensive pig and poultry production is being strongly discouraged. However, few people question the ethics of using antibiotics to curb sickness and losses in pigs when acute disease strikes. Whereas, as yet, consumer attitudes in different countries vary regarding the practice of using antibiotics to promote pig health and production efficiency, the long term goal must be to reduce antibiotics. If consumers are to benefit from this worthy objective and animals are not left to suffer, new strategies will have to be developed. Inevitably, these will be based on environmental enhancement. The identification of best practice will help sustain the cost effective production of wholesome pig meat.

Pig producers will be increasingly regarded as a vital link in the food chain but the ability of the pig farmer to deliver what the consumer wants depends on earlier links in the chain and other environmental issues. Wholesome pig production must be seen alongside demands for less chemical treatment of the land, less pollution of water courses and more trusted raw materials. The interdependence of the cereal producer, pig farmer and retailer has never been greater.

Already, Quality Assurance schemes are driving such strategies. It seems likely that in a global market place, those setting the highest standards eventually will force others to match theirs. If antibiotic use is to decrease, providing better pig environments will become more crucial and consumers will seek increased assurances about such environments.

Consumers are becoming increasingly sensitive regarding aspects of pig welfare in relation to systems of production. This "gut feeling" is being reflected in more exacting demands on some production contracts and has the overall backing of the Pig Welfare Code. Section 5 of the introduction to the UK's Department for Environment, Food and Rural Affairs Pig Welfare Code states:

"No changes should be made to husbandry, equipment or production until the possible effects on animal welfare have been considered. In particular, the possible effect on animal welfare should be considered before installing more complex or elaborate equipment than has previously been used. In general, the greater the restriction imposed on the animal and the greater the complexity of the overall system, the less the animal is able to use its behaviour to modify the effect of unfavourable conditions. Systems involving a high degree of control over the environment should only be installed where conscientious staff skilled in both animal husbandry and the use of the equipment will always be available".

There are also serious concerns for the health of people who look after pigs and the need to safeguard the health of pigs raised for human consumption. Furthermore, there is an increased requirement for the labour force to increase their understanding of pigs and how their welfare needs might be met on the farm. These factors together account for a greater need for pig people to understand and manage pig environments in the future better than in the past.

Familiarity with temperature lift calculations, air velocity measurements, threshold levels of dust and noxious gases is not much help on a Sunday morning to a lone stockman confronted with several pens with fouled lying areas. However, a better understanding of what pigs require, what modern consumers demand, how pigs behave and why they misbehave should help prevent such crises from arising. Although "Pig Environment Problems" is a technical publication, it is not written for the theoretician. It is written for professionals operating at the sharp end who enjoy pigs and want to enhance their comfort with a little help from applied science.

Ethical considerations

Globally, pork is the most important meat product and most consumers purchase pig products on the basis of price. They have little regard for source, production method or processing involved in the sanitisation of their food. Many modern consumers depend on the ability of competing supermarkets to supply a wide choice of visually appealing, wholesome meat products at an affordable price. Pig meat is purchased in a self-perpetuating consumer driven market place. The development of value-added products presents food retailers with a marketing opportunity and they understandably make the most of packaging and presentation which tends to sidetrack any interest in the origins of that product. Food is now sourced from a global production base which encompasses a diverse range of cultures with disparate ethical standards. Shipping food products long distances and presenting them with attractive packaging and labelling tends to erode ethical concerns for most people.

Changing consumer attitudes within affluent societies have resulted in research scientists, veterinary surgeons, building designers, pig farmers and meat processors becoming more aware of ethical issues. Those that neglect ethical issues tend to fall back on power plays, e.g. the "economic necessity" argument to justify their lack of concern; those that have regard for ethics struggle to measure and express their concerns.

It seems reasonable that anyone working in a modern pig industry must have a basic respect for the well-being of the porcine species *per se*. In commercial farming, this is expressed in stewardship and stockmanship which in the absence of other criteria has become the yardstick for the ethical acceptability of different pig keeping systems. The opinions of those not prepared to consume any pig meat whatsoever or any intensively produced pig meat are generally clear-cut. However, there is a marked incidence of duplicity amongst meat eating consumers in our modern and increasingly complex societies.

GENETIC CONSIDERATIONS

Pig environments change over time. The widespread adoption of sow stalls in the 1960s resulted in the selection of "softer" pigs since they were not required to survive in the "free-for-all" regime of group feeding. However, as producers responded to consumer demands, there was a marked change back to group housing triggered by legislation in the United Kingdom. Geneticists responded accordingly and selected breeding stock more likely to thrive in a group housing situation. Furthermore, pigs destined for a sophisticated intensive indoor environment compared to those facing the challenge of the great outdoors clearly have different needs for robustness. Hence, genetic selection can help enhance the comfort of pigs in specific environments.

If pig environments are sub-standard, is it ethically permissible to employ strategies involving genetic manipulation in order to reduce stress within a particular suspect environment? Should not the overall basic behavioural needs and comfort zone of the pig be paramount, compared to an increased dependence on genetic cover-up strategies which help mask shortcomings in the pig environment? The "state of the art" practice of using, e.g. Marker Assisted Selection (MAS) is regarded by some as a potentially materialistic one-sided approach which disregards the environmental needs of the pig. If the genetic integrity of the pig is violated, will not farmers and consumers become increasingly alienated from meat producing pigs? Balancing ethics and molecular genetics with commercial expedience in an increasingly globally competitive market place is a challenge that the traditional pig industry never had to face.

GLOBAL IMPLICATIONS OF ETHICS

The United Kingdom has been a front-runner in the quest to provide better environments for farm animals. However, ethics do not travel particularly well and distance is a great healer for many consumers. Unilateral enforcement of draconian rules regarding pig welfare issues is not the answer. When the UK Government banned sow stalls from 1999 and UK's pig farmers were burdened with extra housing costs, consumers voted with their wallets. The universal nature of the ban encouraged cheaper imports of pig meat from production systems banned in the United Kingdom. In effect, the problems associated with unacceptable pig production systems had been exported and cost conscious consumers asked few questions.

Since those difficult times there has been a marked change in the attitude of some UK consumers because of intense lobbying. Many UK pig farmers were forced out of business but a hard core of producers in desperation formed the British Pig Industry Support Group. Members mounted a high profile campaign which focussed on processors, supermarkets and politicians and took every opportunity to expose the widespread hypocrisy within the fickle market place. The targets of the protestors' actions soon came to realise the short-sightedness of dual standards with respect to pig environment. It became increasingly apparent that premium prices could be obtained for pig meat products differentiated on the basis of production system and supermarkets realised the commercial significance of being seen to conform. Supermarket shopping trolleys contain more than pig meat. Hence, major retailers could not afford to run the risk of missing out on overall sales since they failed to stock a product line favoured by even a small, yet vocal proportion of their customer base. In Denmark, where group housing systems for sows were uncommon, it was soon realised that export market share in the UK would be lost unless production methods took account of the wishes of some consumers. Consequently, the Danes awarded pig farmers premium payments of approximately 5% to those who would commit themselves to producing bacon on systems comparable with those by then obligatory in the UK.

(source: BPEX)

Pig welfare encompasses more than a change in the pig environment. Most countries involved in large scale pig production not only need to be profitable within their home market, but also within a global context. Pig farmers are best able to care for their pigs in an atmosphere of financial security. Upgrading and replacing pig environments demands capital generated from profits. In any event, pig environments are uniquely subjected to challenging physical demands; pig buildings, fixtures and fittings deteriorate rapidly and often compromise pig welfare when maintenance is lacking.

WELFARE ENHANCEMENT

In the United Kingdom, The Royal Society for the Prevention of Cruelty to Animals (RSPCA) in a pro-active move developed the 'RSPCA Welfare Standards For Pigs' which is the vehicle for a unique approval scheme for the rearing, handling, transport and slaughter of pigs. The cornerstone of the scheme, which carries the "Freedom Food" label is the United Kingdom's Pig Welfare Code which is drawn up by the Farm Animal Welfare Council (DEFRA, 2003). However, Freedom Foods goes further in that it assures consumers the pig meat has been produced on systems with enhanced welfare standards.

Like the Pig Welfare Code, the scheme is based on the five freedoms:

- Freedom from hunger and thirst,
- Freedom from discomfort,
- Freedom from pain, injury or disease,
- Freedom to express normal behaviour,
- Freedom from stress.

8 *Pig Environment Problems*

The RSPCA believes that these freedoms would be more likely to be achieved with:

- Caring and responsible planning and management,
- Skilled, knowledgeable and conscientious stockmanship,
- Appropriate environmental design,
- Considerate handling and transport,
- Humane slaughter.

Ultimately, Freedom Foods Ltd aims to establish sister franchise organisations beyond the United Kingdom. They would be required to work to these specific welfare standards so that there is a greater consistency and enhanced ethical standards in pig meat production globally.

Consumer pressures for pig friendly production systems have increasingly become an issue both within the EU and USA. In Europe, discussions are underway regarding the need to include 'non-trade' concerns in future World Trade Organisation (WTO) negotiations. In a bid to curb the tacit importation of pig meat from unfriendly systems of production, the lobby group 'Compassion in World Farming' has formed an unprecedented and unlikely alliance with the European pig industry in order to influence global ethical issues. The WTO Round could jeopardise the future of the EU pig farming if non-trade concerns were ignored and not seen as a future priority for third countries. Furthermore, the General Agreement on Trades and Tariffs prevents the restriction of imports into the EU on the basis of production method. Hence, there is the potential to export the problem of unfriendly pig meat production and consumer choice would be denied. Whereas the EU could press for better production labelling under, e.g. the Freedom Food Scheme, the impact would be limited since it would not address the issue of '"back-door" marketing within processed products and anonymous meat products used in catering. The massive growth in international marketing of processed and oven-ready meat based meals increases the need to adopt global standards for pig production systems.

FOOD CHAIN RESPONSIBILITIES

Pig farmers are increasingly seen as "Food Chain Partners" and within the UK are expected to operate within the spirit of the 1990 Food Act.

Pig Environment Problems 9

This legislation called for greater *traceability* and *due diligence* at all stages of the food chain.

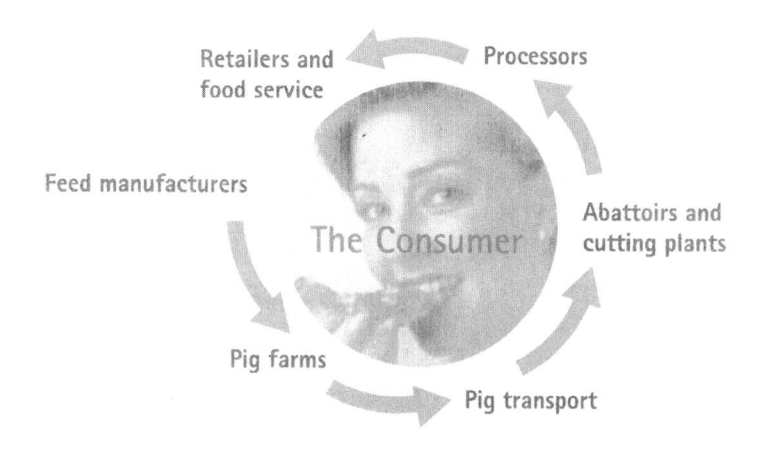

Figure 1.1 Food safety and British pig meat. (source: MLC)

In an era of choice, it has encouraged consumers to question various practices involved in the production of animals for meat. Consumers expect to be able to buy pig meat products at affordable prices but demand safe, wholesome, nutritious food. Unseen practices which boost productivity and profitability on the farm without any tangible benefit for 'spoilt-for-choice' consumers, do little to boost meat consumption. Against this background, the Associated British Pigs Scheme emerged after a tortuous gestation. It provided a single farm assurance scheme for the UK pig industry which served to reassure both supermarket retailers and caterers.

Participating farms are appraised annually by an independent inspector and there is also provision for spot-checks. Samples taken from carcases are tested for antibiotic resistance. The Assured British Pigs Scheme now operates within the Assured Food Standards Scheme which enables well produced pig meat to be labelled with The Quality Standard (Mark of Distinction). This applies to pork, bacon, ham and sausages and conforms to the international standard EN 45011. Several auditing organisations compete to provide certification within this EC approved scheme. The Assured Food Standards Scheme applies to a whole range of food which reaches a specified standard and is denoted by a logo bearing the "Little Red Tractor".

10 Pig Environment Problems

The Quality Standard is very much a food chain initiative and assures consumers that all stages of the production chain conform to high standards regarding:

- Farm assurance,
- Transportation,
- Slaughtering and processing.

Danish service and dry sow accommodation with group housing (Weekly Tribune).

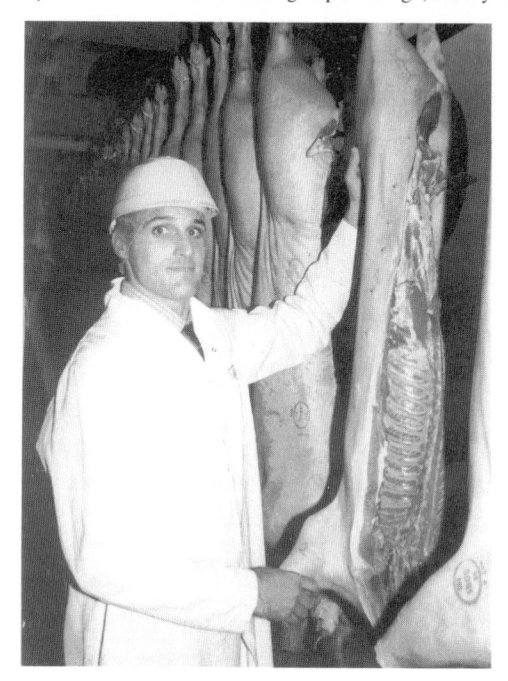

Quality control involves farm assurance, transport, slaughter and processing (George Adams & Sons).

Quality Assurance schemes probably represent the optimum strategy for preventing the modern pig industry becoming ensnared in a food scare trap. A scare is a "disproportionate response to a threat". Affluent, over nourished, "well informed" consumers are regularly force-fed a voluminous diet of perceived threats. Within Europe, North America and Australia, the pig industry, as yet, has not been subjected to a major food scare. However, consumers are regularly subjected to gruesome media coverage involving pig environments, pig transport and processing. Many articulate consumers, particularly those living and remaining in urban environments, confuse their own squeamishness with inadequate pig environments and production practices. Urban dwellers are the major buyers of pig products and trust supermarkets to enforce acceptable quality standards along the food chain. The Quality Standard Mark is recognised by them as a "club badge". Whilst pig environments will never match those occupied by humans living in affluent societies, qualitative monitoring should help safeguard pig producers and help convince most consumers that they are being supplied by caring professionals. Grading pig units with an independently audited "house-keeping score" helps reassure consumers and motivates the pig industry to maintain the highest standards. Independent auditing of the pig environment is an effective means of minimising the risk of food scares and the likelihood of governments introducing panic legislation.

There is every likelihood of more rigorous standards being applied in response to consumer pressures and the need to continue to be regarded as suppliers of a differentiated product. Providing environments which exceed minimum legal requirements will become more widespread within Quality Assurance Schemes.

Modern consumers want more farming and less "pharming" and so there is a need to improve pig environments in order to boost pig health and then demonstrate these benefits to consumers. Furthermore, there is co-ordinated international pressure to reduce the overall use of antibiotics on pig farms in a bid to reduce the likelihood of resistance to antibiotics developing in the human population. The European Commission has outlined proposals to prohibit the use of antibiotics as growth promoting feed additives. The four growth promoters still authorised for use within the EU would have to be phased out by January 2006. A more cautious regime would then operate, any new EU authorisation of antibiotic feed additives would be granted for a 10 year period only. Companies marketing additives authorised under existing regulations would have to apply for re-evaluation and re-authorisation of their products. In America, despite no strong evidence to suggest that humans are developing resistance to antibiotics fed to animals, consumers are becoming increasingly concerned about that possibility. A bill entitled "The Provision of Antibiotics for Human Treatment Act 2002" has been introduced in the USA. The intention is for the early phasing-out of eight classes of antibiotic growth promoter.

Such measures have to be balanced against the indiscriminate use of antibiotics in the treatment of human illness, since this represents the most important cause of antibiotic resistance in people. However, there is limited evidence that resistant microbes originating in farm animals sometimes spill over into the human population. Furthermore, the diverse social cultures within both the EU and global contexts must not be ignored. There seems little point in banning in-feed antibiotics which are unrelated to those used in human medicine whilst, in some countries, people can over the counter purchase antibiotics for their own use without the support of a prescription. Another balancing issue is the need to overcome shortcomings in the animal environment before reaching for pharmaceutical props for food producing farm animals.

Inappropriate environments, in particular, exacerbate respiratory pig diseases. Actinobacillus (App) is one such disease. This bacterial disease is spread amongst pigs which are in close contact with each other (Figure 1.2). It can be spread throughout the herd over a number of weeks and result in severe breathing difficulties and sometimes sudden death. (Figure 1.3).

Shortcomings in the pig environment, particularly on breeder-feeder units where continuous throughput is practised in the grower and finisher accommodation, predispose pigs to this respiratory disease.

Another respiratory disease closely linked with the pig environment is Enzootic pneumonia caused by *Mycoplasma hypopneumoniae* (Figure 1.4).

This disease is endemic in most pig populations throughout the world. When pigs are housed in a good environment, the economic impact of

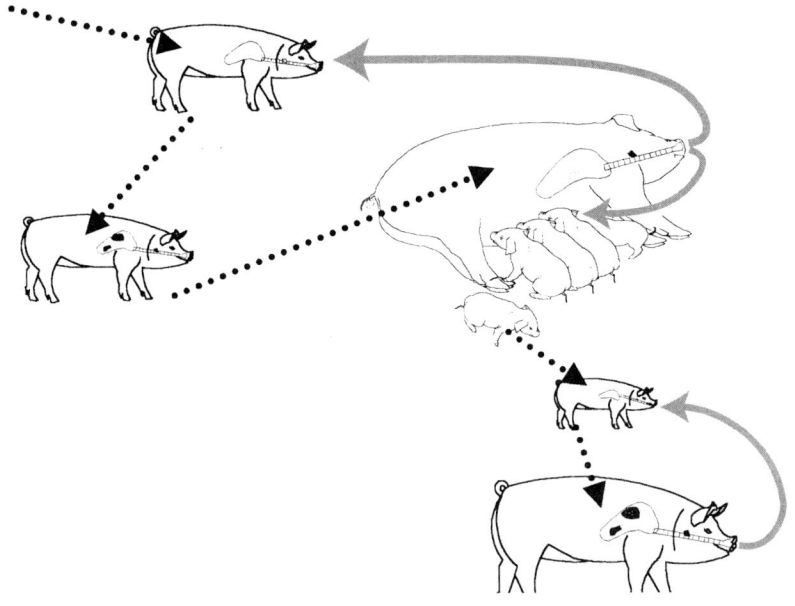

Figure 1.2 Spread of APP in a pig population. (source: Elanco)

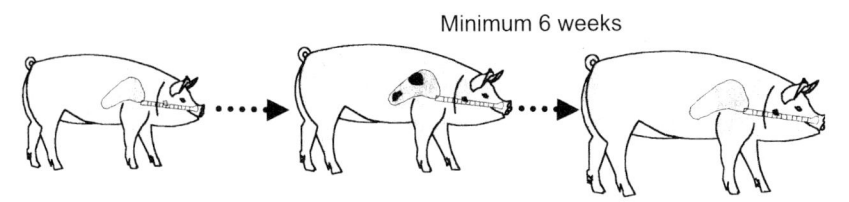

- When naive pigs are infected, APP can be isolated from the upper respiratory tract, the tonsils and the lungs
- From six weeks post infection abscesses and necrotic areas in the lungs are resolved and APP can only be isolated from the tonsils
- As APP is generally spread only by close contact, it can take a number of weeks/months for infection to spread through a population

Figure 1.3 Development of APP in individual animals. (source: Elanco)

the disease is limited and its impact on pig health and enterprise profitability is not so severe. Mycoplasma takes longer to build up and can be spread by airborne droplets both within a herd and from neighbouring pig units. (Figure 1.5)

14 *Pig Environment Problems*

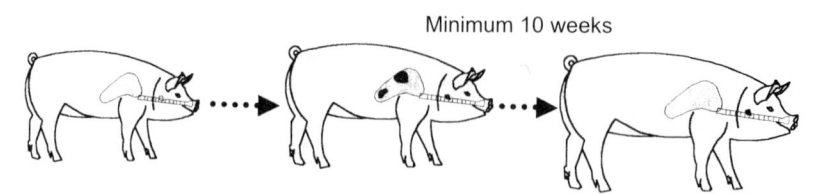

- When naive pigs are infected, Mycoplasma can be isolated from the upper respiratory tract, the tonsils and the lungs
- Abscesses and necrotic areas in the lungs do not develop
- As colonisation occurs on the surface of the trachea and bronchi, immunity takes a long time to develop and a long time to clear infection

Figure 1.4 Develoment of Mycoplasma in individual animals. (source: Elanco)

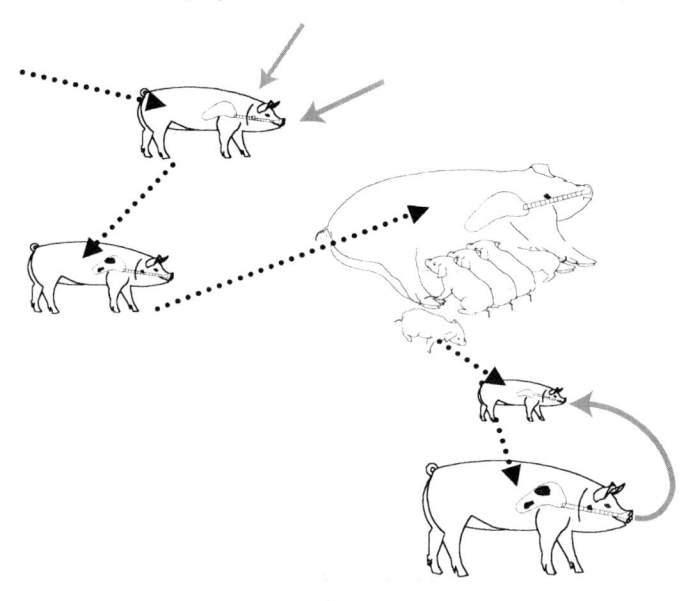

- Infected grower finishers and replacement gilts excrete Mycoplasma into the environment.
- These airborne Mycoplasma from this farm and neighbouring farms are the major source of infection for naive pigs

Figure 1.5 Spread of Mycoplasma in a pig population. (source: Elanco)

Is it unethical to predispose pigs to illness because of shortcomings in the production system? Many breeder-feeders with less than 200 sows are burdened with pig unit layouts which do not facilitate all-in/all-out production and this can permanently compromise pig health. Masking the impact of sub-standard environments and management practices with blanket medication is an unacceptable practice. However, partial depopulation of the pig unit, accompanied by a period of "blitz medication", provides a possible escape route from this health trap. The strategy is outlined in Figure 1.6.

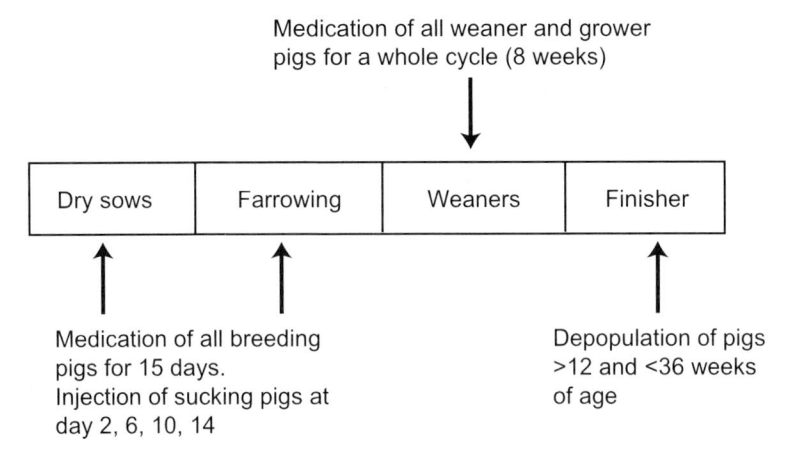

```
              Medication of all weaner and grower
              pigs for a whole cycle (8 weeks)
                           |
                           ↓
┌───────────┬───────────┬───────────┬───────────┐
│ Dry sows  │ Farrowing │  Weaners  │  Finisher │
└───────────┴───────────┴───────────┴───────────┘
      ↑           ↑                       ↑
Medication of all breeding        Depopulation of pigs
pigs for 15 days.                 >12 and <36 weeks
Injection of sucking pigs at      of age
day 2, 6, 10, 14
```

Figure 1.6 Medication strategy.

Periodic emptying and cleaning of pig buildings provides an opportunity to appraise and upgrade the pig environment. The benefit of partial depopulation tends to be short-lived if pigs continue to be subjected to wet floors, temperature fluctuation, draughts, inadequate ventilation and inappropriate stocking density. When partial depopulation is combined with environment enhancement, there can be a lasting impact on pig well-being and profitability. It can be a useful escape from the trap of routine antibiotic use and reduce the likelihood of any build-up of antibiotic residues.

In the UK the Veterinary Medicines Directorate (VMD) is responsible for surveillance for veterinary residues in meat. The Directorate undertakes statutory surveillance in accordance with the provisions set out in the EU Council Directive 96/23/EC. Annex IV of this directive defines the number of samples that EU member states are required to test. This is related to each country's forecast for the likely number of slaughterings in any particular year. In the UK the State Veterinary Service is responsible for the collection of samples from feed and from live pigs on farms. Responsibility for collection of samples at the abattoir rests with The Meat Hygiene Service. A key aspect of the scheme is the reporting of results to abattoir operators within 10 working days of the end of each month.

Non-statutory surveillance for veterinary residues is also undertaken within the UK. It targets foods and residues not included in the Statutory Surveillance Programme. Using data from the National Food Survey, the most popular preparations of meat and meat products are tested. This indicates sampling from a wide range of retail outlets using a protocol that ensures traceability back to the supplying farm. Additionally, samples of imported raw meat are taken at Border Inspection Posts. In effect, there is a very comprehensive on-going programme which ensures that veterinary residues in pig meat products are monitored. Results are published in detail in the Veterinary Medicines Directorate annual report.

In Great Britain in the year 2000 a total of 2,221 samples were taken from all red meats. Only 0.14% showed any signs of traceable residues and only 18 samples (0.07%) contained residues above the Maximum Recommended Limit. No evidence of synthetic hormones or beta-agonists were found in any live or slaughtered pigs but the main residues found were sulphonamides. Additionally in 2001, 1,320 samples were scheduled to be collected and 7,726 analyses undertaken within the Non-Statutory Surveillance Scheme. Data from the Danish Veterinary Services for the year 2000 also indicates the great care that is being taken to avoid pig meat being contaminated with medication, only 3 out of 20,474 samples tested positive for antibiotics. Table 1.1 shows that the year 2000 figure is indicative of the long term trend in Denmark.

Table 1.1 Residue surveillance - antibiotics. Positive samples from 1995-2002.

1995	1996	1997	1998	1999	2000	2001	2002
0.02%	0.02%	0.03%	0.01%	0.02%	0.01%	0.005%	0.02%

(source: Danske Slagterier)

Although there is some annual variation, detection of antibiotic residues is consistently low in Denmark.

An effective means of reducing antibiotic residues in meat producing animals is to feed less antibiotics. Historically, antibiotics have, in some instances, been used in pig rations to help reduce the shortcomings of inadequate housing and poor management. Such a philosophy is not consistent with both the perception and reality of wholesomeness. Sweden has set the pace in banning the routine use of antibiotic feed additives. The Danish Slaughter Houses Association (Danske Slagterier) has also been pro-active in avoiding dependence on antibiotic growth promoters and by the year 2000 had banned their regular use in pig rations. This has inevitably increased pressure on the European Commission to bring about such changes in other countries. In 1996, eleven anti-microbial products had EU licences for routine inclusion in pig feeds. By 1999 the number had fallen to four in an era of increased surveillance. The dilemma for the European Commission and EU pig keepers has parallels with the global pig welfare scenario. Danish efforts to ensure wholesomeness of pigmeat are summarised in Figure 1.7.

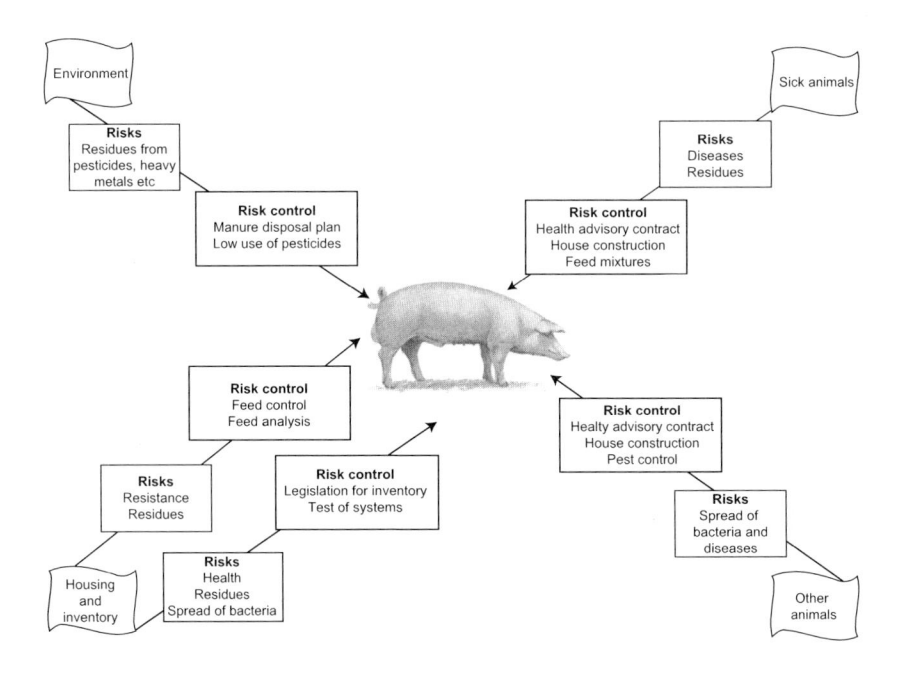

Figure 1.7 Herd management for wholesome pigmeat production (source: Danske Slagterier).

The Scandinavian model involves a ban on all digestive enhancers irrespective of the scientific evidence. Furthermore, in Scandinavia, there are great pressures to restrict and reduce the use of therapeutic antibiotics. It has become a political issue which, as yet, cannot be supported on a scientific basis. This political dimension was confirmed in 1999 in a statement from the USA's National Research Council which declared:

"Until more accurate data on animal antibiotic use, patterns and rates of resistance transfer to humans, occurrence of actual disease emergence

and mechanisms of resistance are available, actions aimed at regulating antibiotics cannot be implemented through a science driven, well validated, justified process".

Perception is the driver of the modern market place. In Denmark between 1996 and 2000, the overall level of drug usage within the Danish pig industry halved (Figure 1.8).

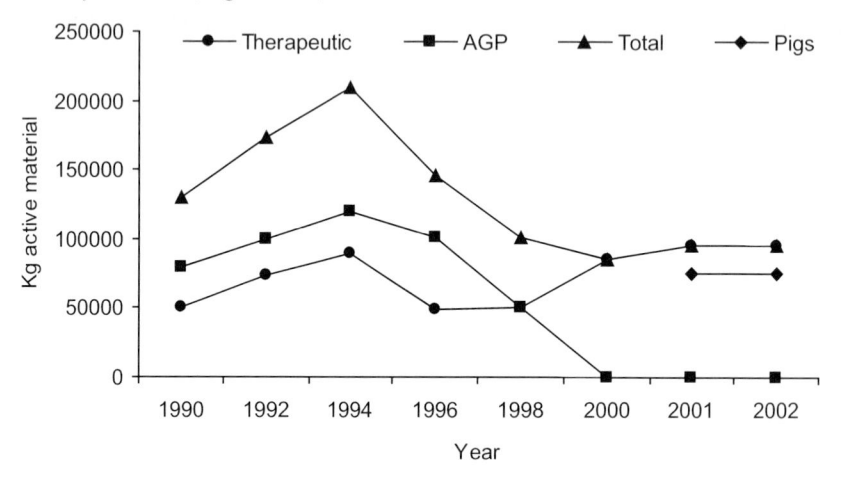

Figure 1.8 Use of antibiotics in Danish livestock production (source: Danske Slagterier)

This reduction arose mainly from the total ban on antibiotic growth promoters. During the same period, total pig production in Denmark increased by almost 10%, hence antibiotic usage per pig actually declined by 60%.

Following Denmark's exclusion of antibiotic growth promoters from weaner rations in January 2000, there was an increase in the use of *therapeutic* antibiotics. The Danes regarded this as a "blip" reflecting the steep management learning curve pig producers had to ascend and that the increase would be short-lived.

It would be unethical to administer medicines to pigs if there were any likelihood of meat eating consumers being harmed. Studies are undertaken to ensure how quickly residues of all medicines are eliminated from the pig. Minimum Residue Limits (MRLs) are determined and these represent the maximum permissible levels of a drug's active ingredient remaining in pigs destined for human consumption. A huge safety margin is built into the MRL figure so that consumers are reassured about meat safety.

Various antibiotics take different time intervals before falling below the MRL in the slaughter pig. Hence a specific "withdrawal period" is enforced for every drug administered. This ensures that there is sufficient time between administration of the medicine and slaughter of the pig, to ensure that the Minimum Residual Level would easily be achieved. The Animals & Fresh Meat (Examination for Residues) Regulations 1988 requires pig farmers to keep records of medicines administered so that withdrawal periods can be checked. Antibiotics should only be administered to pigs for the best possible reasons and in a regime with which consumers feel comfortable.

There seems to be questionable logic in the ethics of antibiotic use. Ironically some consumers readily administer antibiotics to combat their own maladies and those of their family, yet feel uncomfortable with the concept of sick pigs receiving medicine. Humans living in super-hygienic environments suffer from bacterial and viral diseases and pigs, even when kept in the cleanest environment, sometimes also succumb to health challenges. It seems ethical, therefore, to grant indisposed pigs the same "rights" as humans.

The perception that untested drugs are regularly and effortlessly brought to the market place is a myth. Development of new medicinal products from scratch is a lengthy and expensive process. Once a drug has been identified as having the potential to safeguard pig health, before it can be brought to the market place, it has to comply with an array of demanding safeguards. These include:

- laboratory studies regarding pharmacology and mode of action
- toxicity tests on the pig (acute and long term)
- toxicity tests on the stockperson administering the medicine
- toxicity tests on the consumer of products from treated pigs
- toxicity tests which might be imposed on the environment
- analytical methods have to be developed to detect residues
- species specific dose measuring trials to ascertain appropriate and effective dosage levels
- establishment of Maximum Residue Limit for the active ingredient
- depletion studies to determine safe withdrawal periods
- scaling up of laboratory preparation to commercial production levels
- formulation and stability trials
- field trials to test the product under commercial farm conditions

Despite the concerns of some consumers, within the World Health Organisation, as yet, the thinking is that decisions regarding antibiotic usage should be based on scientific evidence. Their caveat is that any digestive enhancers used in animal feeds should not be the same type as those used in human therapy or likely to interfere with such products. Meanwhile, the responsible therapeutic use of antibiotics is generally likely to continue on pig farms and inevitably there will be even more effective surveillance of all antibiotic use.

Coping strategies on pig farms

The Responsible Use of Medicines in Agriculture Alliance (RUMA) comprises a group of organisations representing agricultural, veterinary, pharmaceutical and retail interests. It was set up in the United Kingdom to address concerns about cross-over of resistant bacteria from livestock to the human population. RUMA has set out comprehensive guidelines for the responsible use of anti-microbials in pig production,

Pig farmers are advised to:

- Regard therapeutic anti-microbial products as complementing good management, vaccination programmes and site hygiene.

- Initiate medication with a medicine subject to a veterinary prescription only with formal veterinary approval.

- In the case of in-feed medication this will be provided by a Medicated Feedingstuff (MFS) Prescription.

- Ensure that accurate information is given to the veterinary surgeon in order that the correct dosage can be calculated for the animals concerned, and ensure that clear instructions for dosage and administration are obtained and passed on where necessary to the staff responsible.

- Always complete the course of treatment at the correct dosage. Ensure that the dosage is carefully administered in an effective manner.

- Accurately record the identity of the animals medicated, the batch number, amount and expiry of the medicine used, the withdrawal period required and the date and time the medication was completed.

- For in-feed or in-water medication, ensure that the end of medications is accurately determined by cleaning the header tank or feed bin as appropriate.

- For any medicines used, appropriate information should be kept on file – for example, product data sheets, package inserts or safety data sheets as available.

- Report to the veterinary surgeon (or direct to the Veterinary Medicines Directorate) any suspicion of an adverse reaction to the medicine in either the treated animals or farm staff having contact with the medicine. This should include any unusual failure to respond to medication. A record of the adverse reaction should also be kept on the farm: either a copy of the VMD adverse reaction form or a note in the medicines record book.

- Ensure that the appropriate withdrawal period is complied with prior to slaughter of the treated animals for human consumption. In general the withdrawal time required is specified on the Medicated Feedingstuff Prescription or on the label of the medicine.

- Co-operate with Farm Assurance schemes which monitor anti-microbial usage, medication documentation and withdrawal period compliance. However, such schemes should not constrain the attending veterinary surgeon from preventing suffering in the animals under their care.
- With your veterinary surgeon track anti-microbial usage taking account of the potency of various products. The simplest approach is to record the number of kg of animal treated per day as a proportion of the total kg of animal at risk.

- Ensure that different medicines are only given at the same time with the specific approval of the veterinary surgeon because adverse interactions can occur.

- Maintain a medicines log book on farm together with copies of relevant regulations and Codes of Practice.

The medical profession also has responsibilities regarding the widespread prescribing of antibiotics. There is now some optimism that in the not too distant future when disease first strikes, doctors will be able to use DNA profiling to identify which particular strain of bacteria are causing an infection. This would then facilitate use of a narrow spectrum antibiotic, make broad spectrum antibiotics redundant and so reduce the incentive for the development of bacterial resistance.

If, say by 2010, the routine use of anti-microbials within the EU pig industry were phased out in line with the Scandinavian precedent, what would happen if countries outside the EU continued to use existing technology and grasp new technologies? Very likely the EU would be flooded by imports of lower priced untraceable pig meat and consumers would become the "victims". The market would be awash with oven-ready meals comprising "de-animalised meat" of questionable origin offered to consumers in attractive user-friendly packaging. Whatever the outcome, the need to provide the modern pig with a cleaner, healthier environment is sacrosanct.

Other consumer issues

Some "enlightened" consumers have linked into the organic food chain in order to maximise on wholesomeness. Many are concerned and committed meat eaters who are prepared to pay a premium for food marketed with certain safeguards regarding the use of additives and processing aids. However, internationally respected certification schemes for organic production permit up to 20% of the daily dry matter intake of pigs and other meat-producing animals to be from non-organic sources. It seems reasonable for a buyer of organic ham to assume that such premium products are natural and contain no added chemicals. The reality is that organic ham is laced with both sodium ascorbate and sodium nitrite. The former is an antioxidant and sodium nitrite is a preservative, which some scientists believe is carcinogenic and may bring about hypersensitivity in children. As pig production and pig meat processing becomes increasingly global and added value products are more aggressively marketed, the need for improvements and standardisation of the pig environment is likely to increase.

There will be increasing demands on the designers and managers of buildings used for mainstream pig meat production to minimise the risk of food borne disease. Such diseases arise following the consumption of food contaminated with micro-organisms and toxins and can lead to diarrhoea, vomiting and abdominal pain in its victims.

Salmonella is a food borne disease and infections in the human population are widespread and persistent throughout the world. However, the reported incidence in some European countries has been up to 80 times that in the USA. Whilst such variations would be a reflection of differences in livestock housing systems, different approaches to food processing and handling could also be implicated. The awareness and reporting of more health problems in any food animals does not necessarily reflect an increased incidence of disease. Those with the foresight for increased vigilance and improved recording and reporting systems are sometimes misrepresented, transparency can have a cost.

Comparative Danish data for salmonella outbreaks in broilers, pork and eggs are shown in Figure 1.9.

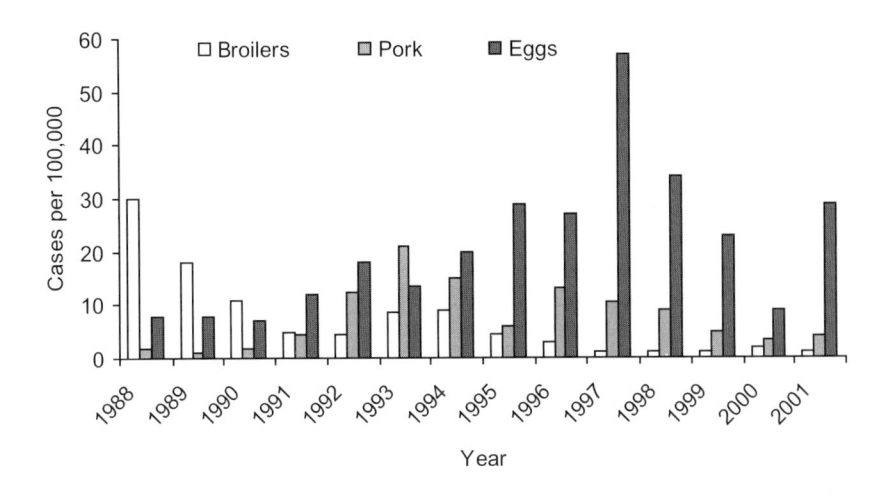

Figure 1.9 Main sources of human salmonellosis in Denmark, 1988-2001 (source: Danske Slagterier)

The histogram indicates a vast decline in salmonella infections arising from broilers and generally higher levels associated with egg production compared to pork production.

In Denmark, a national programme of salmonella control has been underway since 1995. The Danes have adopted a food chain approach to their monitoring protocol. Initially, fresh meat from pig herds was sampled on a monthly basis. Since January 2001 the sensitivity of the monitoring has been increased by analysing whole carcasses on the production line in an attempt to meet more demanding targets. When the new testing regime was first introduced, initially higher levels of salmonella were detected but, within 3 months, a downward trend was noted.

Danish pig units are given a score indicating the degree of incidence of salmonella (Table 1.2).

Table 1.2. Salmonella control in Danish pig herds (July 2001)

Level	Number of herds	%
1 (Reactor Index <40)	13,021	96.3
2 (Reactor Index 40-69)	367	2.7
3 (Reactor Index >70)	130	1.0

Source: The Annual Report on Zoonoses (Danish Zoonosis Centre) 2000

Herds within the Level 1 grouping are deemed to be acceptable and no management changes are recommended. Those with higher scores, i.e. 40 and above are required to make significant changes if they are to remain a link in the pig meat chain. Table 1.2 indicates that over 96% of Danish herds were classified as 'safe' by July 01.

In England and Wales, the Public Health Laboratory Service (PHLS) is responsible for monitoring salmonella food poisoning in humans. In Scotland the same role is undertaken by The Scottish Centre for Infection and Animal Health. Figure 1.10 logs the incidence of salmonella poisoning in humans between 1981 and the year 2000.

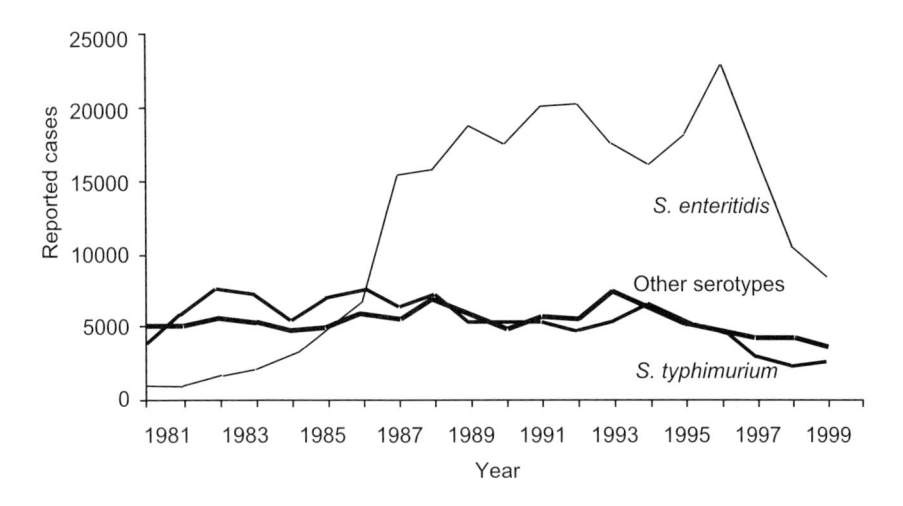

Figure 1.10 Salmonella in humans in England and Wales, 1981-2000 (source: Public Health Laboratory Service)

Pork and bacon have rarely been linked with salmonella food poisoning in PHLS investigations. Figure 1.10 indicates that *Salmonella enteritidis*, which is almost exclusively found in poultry, is the main form of salmonella infection of food in humans. In the year 2000, almost 60% of salmonella problems in humans were attributed to *S. enteritidis*. In England and Wales, salmonella infections in humans peaked in the mid-nineties. However, practices were changed and, by the end of the decade, levels were at a 14 year low. Nevertheless, human sickness resulting from salmonella infections are thought to cost the United Kingdom around £15m/year.

After a consultation exercise involving the Food Standard Agency, National Pig Association and Pig Veterinary Society at the end of 2000, the UK government issued a voluntary "Code of Practice for the Prevention and Control of Salmonella on Pig Farms". Tables 1.3, 1.4 and 1.5 highlight some key recommendations (Annex 1-3) from that publication. A core recommendation is the need to ensure cleanliness of the pig environment and minimise the likelihood of faecal contamination to the skin.

26 Pig Environment Problems

Table 1.3 Strategies for reducing the incidence of salmonella on pig farms

Annex 1: Visitors Book

The following headings are recommended:

- Date
- Name of Visitor
- Company name/address
- Purpose of visit
- Date of last contact with pigs
- Address of last contact with pigs
- Time of arrival
- Time of departure

Table 1.4
Annex 3: Check list for preparation of a detailed plan for cleaning and disinfection of pig units

PREPARATION:	CHECK ✓
Note depopulation date and prepare a plan	
Consult COSHH assessments	
Ensure rodent controls are effective	
List items for repair and maintenance and order replacements	
Ensure cleaning equipment, disinfectant (Defra approved) available	
Ensure competent staff available	
Ensure other animals will not be contaminated	
Run down feed supply	

AT DEPOPULATION:	CHECK ✓
Remove all livestock from the building	
Check rodent control effective/intensify as necessary	
Apply insect control measures as necessary	

CLEANING AND WASHING:	CHECK ✓

Clean out manure, bedding, dust, waste etc.
Take all moveable equipment outside, clean and wash
DANGER - disconnect electrical equipment as necessary
Drain, flush, clean water system, dismantle as necessary
Clean feed troughs thoroughly, feed areas, bins, hoppers etc
Clean ancillary rooms, fans, storage areas and rest rooms
Clean bins used for waste material, boot dips
Pressure wash the building, pens, other areas to remove remaining dirt
Dispose of all waste safely
Ensure that all cleaning equipment is cleaned and disinfected
Carry out repairs and maintenance

APPLYING DISINFECTANT:	CHECK ✓

Ensure the building is dry
Follow label instructions and COSHH
Apply Defra approved disinfectants at approved rates to:
- the building (including the water system)
- moveable equipment and reassemble
- all ancillary and common areas
- feed storage areas, bins, hoppers

BEFORE RE-STOCKING:	CHECK ✓

Replace rodent bait
Check no areas overlooked and equipment is functioning
Ensure route of entry for new stock has been cleaned and disinfected

In 2002 within the United Kingdom, the Zoonosis Action Plan (ZAP) was launched. This was a co-operative venture involving the Meat and Livestock Commission (MLC), DEFRA (Department for Environment, Food and Rural Affairs) and the Food Standards Agency. The project

28 Pig Environment Problems

Table 1.5 Salmonella control

Annex 2: Salmonella control – a summary

Control point	Keeping salmonella out	Controlling the spread
Unit	For new units – locate well away from other farms, in particular pig farms and landfill sites. Keep clean and tidy. Perimeter fence/information signs. Parking for vehicles. Provide washing/disinfection facilities/footbaths. Clean and disinfect regularly	Keep clean and tidy. Provide washing/disinfection facilities/footbaths. Clean and disinfect regularly.
Stock	Introduce a salmonella monitoring programme. Operate all in/all out system. Purchase stock from reliable source. Isolate/quarantine purchased stock.	Operate all in/all out system. Keep pigs clean. Segregate groups. Isolate sick pigs/infected group.
Staff	Train and inform. Keep "work clothes" on site and clean and disinfect regularly.	Keep "work clothes" on site and clean and disinfect regularly.
Pest Control	Effective control programme	Check controls effective
Visitors	Restrict entry. Visitors book. Provide clean protective clothing	Provide clean protective clothing
Feed	Reliable source/salmonella tested. Secure, clean storage away from pigs. Mixing/milling away from pigs.	Check for signs of contamination. Check storage secure
Bedding	Clean source, not contaminated	As for feed
Water	Mains or tested source	Check for signs of contamination. Enclosed water system
Animal waste	Careful disposal away from site	Store slurry for at least 4 weeks. Cover and compost manure. Spread on arable land but away from near-harvest crops.
Equipment	Do not share equipment. Clean and disinfect regularly	Clean between sections of the farm. Clean and disinfect regularly

Source: Department of Food, Environment and Rural Affairs

involved testing for salmonella antibodies in pig meat. All abattoirs involved in the British Quality Assured Pig Scheme participated. All countries with salmonella monitoring schemes are keen for governments to implement testing of imported pig meat. It seems reasonable for those farmers producing pigs from a "salmonella safe" environment to be allowed to differentiate their product in the market place

Within the ZAP scheme, any farms where there was a high prevalence of salmonella were identified. These producers were offered remedial advice and, with the veterinary surgeon, developed an action plan to reduce salmonella levels. In 2001 a Veterinary Laboratories Agency study of the epidemiology of salmonella was initiated. The objective was to identify risk factors and try to define effective control measures. Some of these factors could be environment specific, hence MLC are also undertaking a detailed study of salmonella incidence in finishing buildings with different flooring systems. Wet feed has a low incidence of Salmonella contamination on account of its low pH. Pig feed can, however, be a source of salmonella contamination: pelleted feeds are virtually negative for salmonella, whereas meals are around 2% positive. Pig nutritionists are increasingly mindful that the broiler industry has reduced salmonella levels in the poultry environment. Broiler feed is routinely heated above 80°C and treated with acids to reduce contamination risk.

Research has indicated that in particular, organic acids can markedly suppress salmonella levels in dry material such as pig feed.

Table 1.6 Effect of WHEAT TREET, a liquid salmonella inhibitor on *Salmonella enteritidis*

Time (hrs)	Salmonella organisms/g Control	Inhibitor (3kg/tonne)
4	27	23
8	17	9
24	16	5
48	11	<1

Source: Kemin Europa N.V.

Table 1.6 indicates that, when whole wheat was treated with a commercially available organic acid, a substantial fall in the level of *S. enteritidis* was noted. Hence this approach could become a vehicle for driving pig feed into microbiologically safer country. Organic acids are one of a group of components of feed known as *nutricines* and they exert a beneficial effect upon health and metabolism. They act as a link between health and nutrition and, in the absence of antibiotics, could become increasingly important in providing a generally healthier pig environment.

Responsibilities concerning pig farm staff

Whereas those involved in the pig industry have a moral duty and commercial need to safeguard the health of their pigs within any particular environment, there are extra responsibilities towards people. In the UK this is backed by tough legislation which could lead to a pig farmer facing the prospect of a fine of up to £20,000 or six months imprisonment if disregarded. Since the introduction of the 1992 Management of Health and Safety at Work Regulations, *every* employer has been required to make a "suitable and sufficient" assessment of:

a. The risk to the health and safety of his or her employees,
b. The risk to the health and safety of persons not in his employment arising out of, or in connection with his undertaking.

This information must then be used to enable the employer to identify measures needed to comply with requirements and prohibitions and the employer is also required to keep abreast of any changes. The significant findings of the assessment have to be recorded if the place of work has five or more employees. Hence, those running large pig units have extra responsibilities regarding the working environment. There are five key steps to carrying out Risk Assessment:

Step 1 - Look for the hazards
Step 2 - Decide who might be harmed and how
Step 3 - Evaluate the risks arising from the hazards and decide whether existing precautions are adequate or more should be done
Step 4 - Record your findings
Step 5 - Review your assessment from time to time and revise it if necessary

Pig Farmers must consider whether a *hazard is significant* and whether it has been covered by satisfactory precautions so that the *risk is small*. Table 1.7 outlines a system for scoring both *Probability* and *Severity* of a potential hazard, the lower the score, the greater the need to improve the working environment. Table 1.8 gives some guidance on Action Criteria.

Table 1.7 Risk assessment action criteria: severity and probability scores

Probability	Value	Severity	Value
Probable – likely to occur immediately or shortly	1	Catastrophic – imminent danger exists. Hazard capable of causing death and illness on a wide scale	1
Reasonably probable – Probably will occur in time	2	Critical – hazard can result in serious illness, severe injury, property and equipment damage	2
Remote – may occur in time	3	Marginal – hazard can cause illness, injury or equipment damage but results would not be expected to be serious	3
Extremely remote – unlikely to occur	4	Negligible – hazard will not result in serious injury or illness. Remote possibility of damage being beyond first-aid case.	4

Table 1.8 Hazard rating index

Score	Action
1	Suspend operation and take immediate action
2-3	Second priority: respond as soon as practicably possible
4-6	Third priority: respond within two months or before end of season (if seasonal activity)
8-12	Fourth priority: respond within six months or before the beginning of the new season (if seasonal activity)
16	No action necessary

If the product of the scores for Probability and Severity rating is as low as 1, the offending activity in that environment must be stopped and emergency measures implemented. Conversely, if the Probability x Severity Score is as high as 16, then no action is necessary.

32   *Pig Environment Problems*

A practical example of Risk Assessment in respect of the care and handling of live pigs is provided in Table 1.9.

Table 1.9 Risk assessment record

Subject:	Care and handling of live pigs
Impact:	Accidents/ill health of people
Susceptible individuals:	All pig unit staff, some visitors

Evaluation Summary

Hazard	Severity	Rating Probability	S x P score
Physical contact: attack, bites and crushing	1	2	2
Zoonoses: disease transference	3	2	6
Dust: irritation, respiratory stress	1	2	2
Routine administration of medication to healthy and sick pigs: oral poisoning, self-injection	3	2	6
Routine mucking-out of odorous soiled bedding: physical contact, nausea	3	2	6
Manual handling: physical injuries	2	2	4
Machinery and Equipment: physical injuries	3	1	3

In this instance, dust has been identified as a serious short-coming in a particularly poor pig environment. Sooner or later, very likely good luck would run out and someone involved in care of the pigs would be likely to suffer a serious respiratory health problem. Urgent remedial action would be required and changes to the pig environment must be made as soon as practicably possible.

In order to reduce the likelihood of any aspect of the environment becoming a health problem throughout the pig unit a *Safe Working Practices Strategy* must be developed. A typical strategy for minimising dust problems would be:

- Identification of high risk operations and locations, e.g. bedding, weighing, mucking-out, brush work, feeding, pig moving.

- Insistence that face masks must be worn in high risk situations.

- Abiding by Control of Substances Hazardous to Health (COSHH) regulations.

- Using air extraction equipment where provided.

A major study in Scotland involved monitoring the respiratory health of pig farm workers and their immunological response to airborne contaminants (Crook *et al*, 1991). A total of 29 workers completed a questionnaire and 24 of these provided blood samples for the measurement of specific Ig E and Ig G antibodies to pig and feed related environmental contaminants. The working environments involved 20 pig buildings where total airborne dust levels varied from 1.66 to 21.04 mg/m^3 and ammonia varied from 1.50 to 13.23 ppm. This wide variation reflected seasonal effects and the different feeding systems underway.

Microbiological monitoring indicated a predominance of gram-positive cocci bacteria. *Staphylococcus lentus, Aerococcus viridans, Staphylococcus hominis* and *Micrococcus lylae* were identified in more than 75% of air samples tested. Most of the bacteria isolated were those associated with mammalian skin and pig feed. In some instances a range of fungal species was also found. Aerodynamic particle size diameter of the bacteria collected was generally 6μm or above, hence most of the bacteria were deposited in the upper respiratory tract rather than the alveolar region.

Twenty three out of 29 employees reported work related respiratory symptoms. A summary of the findings is given in Figure 1.11.

Twenty out of 29 employees complained of nasal and eye irritation, five of these workers also suffered from coughing and nine experienced wheeze and chest tightness problems. Eight of the pig farm workers believed that their symptoms were associated with pig weighing, whereas six blamed

34 Pig Environment Problems

milling and mixing. Six workers reported suffering from headaches, particularly when pig buildings carried an odour of ammonia. Medical tests indicated that three of the workers had impaired lung function, the remainder were within the normality range. This Scottish research clearly indicated that shortcomings in the working environment can impair the health of pig farm workers.

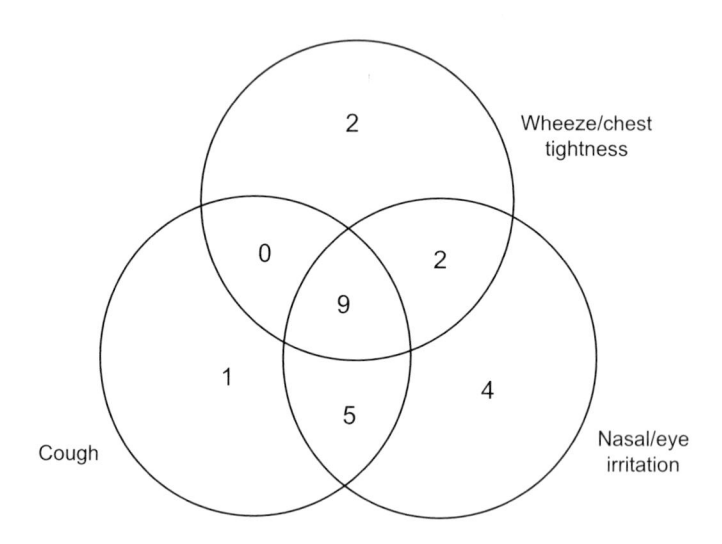

Figure 1.11 Distribution of work-related respiratory symptoms amongst 29 pig farm workers (source: Crook *et al.* (1991))

Good pig environments do not arise by chance. They must be planned, established, monitored, managed and up-graded as required. Respiratory diseases are exacerbated when pigs and people are subjected to dusty environments, particularly when ammonia levels are high. Rhinitis and enzootic pneumonia are more likely to arise in pigs when pathogens are present within an environment where the ammonia concentration exceeds 10ppm at pig level. In such environments, those caring for the pigs are more likely to suffer from respiratory ailments such as bronchitis and hypersensitivity which is far from conducive to recruiting and retaining quality labour.

Global environmental responsibilities

Ammonia arises because surplus nitrogen in the feed is eliminated by the pig and ultimately is re-cycled through the soil profile in the form of nitrates. Not all of this is absorbed by plants and some finds its way into water courses. Under the E.C. Nitrate Directive (91/676), all EU Member States were required to set-up a voluntary code of practice to safeguard ground water, streams and rivers from nitrate pollution. Additionally the UK Government has brought in mandatory legislation which targets land known as Nitrate Vulnerable Zones (NVZs) which have been identified as land areas draining directly into water courses and where the nitrate concentration in the water exceeds or could exceed 50mg per litre.

Depending on the genetics, sex, age of pig and productive process, only 20-40% of the nitrogen fed in protein to pigs is retained. The surplus is flushed through the kidneys and is excreted as urea in urine and undigested residues in faeces; most of this is converted to ammonia. The need to flush away the surplus with drinking water reduces when lower protein diets are fed to such an extent that, because of this reduced water intake, the overall volume of slurry can be reduced by around 30%.

Trials undertaken at ADAS Terrington involved feeding growing and finishing pigs:

A control diet : -   20.6% crude protein for growers
                -    18.7% crude protein for finishers
or

A low protein diet: -   15.2% crude protein for growers
                -    13.0% crude protein for finishers

Levels of the amino acids lysine, threonine, tryptophan and methionine were maintained at the same level in the low protein diet as in the control (Kay and Leigh, 1977).

Figure 1.12 indicates that the amount of nitrogen excreted can be lowered by around 40% without detriment to feed efficiency and pig performance. The research confirmed that: *"Low protein diets offer a practical and economical way to reduce the amount of nitrogen excreted by pigs. They also reduce the emissions of gases harmful to stock people, animals and the environment"*.

36 Pig Environment Problems

Ammonia emissions (g/24hr/24 pigs)

Ammonia concentration (ppm)

□ Commercial diet
■ Low protein diet

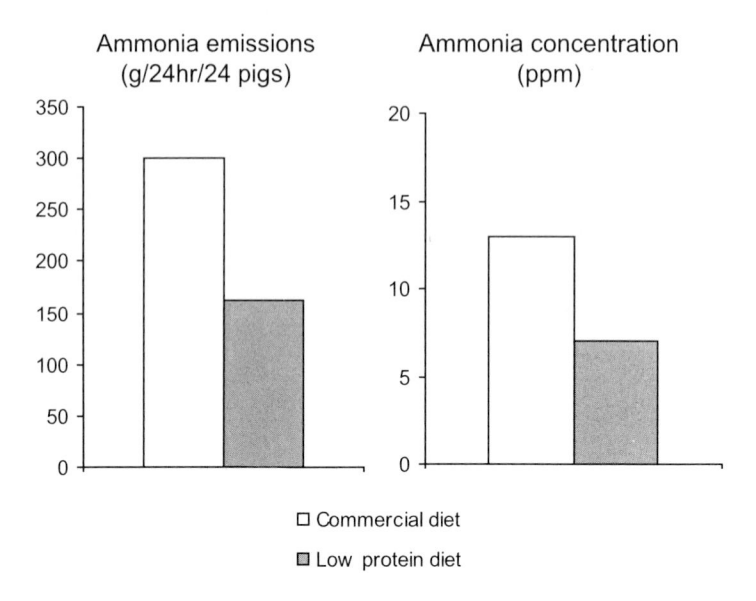

Figure 1.12 Source: RM Kay and PM Leigh (1997)

Pollution prevention and the safeguarding of amenity are generally being strengthened by demanding legislation which traditional pig keepers would never have imagined. The E.U. Council Directive 96/61 sets out new rules for Integrated Pollution Prevention and Control (IPPC). At present in the United Kingdom, these measures apply to larger pig units, larger poultry units and a wide range of other industries. The IPPC regulations aim to minimise damage to the environment ideally by *preventing*, or where not practical, *reducing* emissions to air, water and land and promoting the careful use of resources such as water and electricity.

Designers of pig buildings and environment control systems will increasingly have to comply with global responsibilities as outlined in IPPC regulations. The problem of wasteful use of scarce energy resources must be addressed by the pig industry and those who choose to waste energy will have to bear the financial penalties as decreed in the Climate Change Levy. Since 2001, in the UK, both pig and poultry farmers have been able to register for an 80% rebate on this levy. In return they have to deliver agreed improvement in energy efficiency over a ten-year period.

Green Farm Energy, Denmark where technology uses pig slurry as a source of power generation (Weekly Tribune).

IPPC demands have, in the UK, become enshrined in national law within the Pollution Prevention and Control (England and Wales) Regulations 2000. By January 2007, all pig units with more than 750 breeding sows or 2,000 finishing pigs over 30kg live weight will have to apply for an IPPC permit. Prior to 2007, any existing large pig units where there is *expansion* or *substantial change* will also have to apply for a permit. Furthermore, any new pig unit built after the IPPC rules were introduced but built before 2007, will also be subjected to these demanding EU regulations, if the pig place threshold were exceeded (DEFRA. 2004).

Large units will be subjected to Integrated Pollution Prevention and Control Regulations (BPEX/Rattlerow Farms).

In order to obtain a permit, pig farmers will be required to convince the Environment Agency that the pig unit conforms to IPPC requirements. In particular, the Environment Agency must be convinced that:

- all preventative measures against pollution have been taken
- no significant level of pollution will be caused
- waste disposal has been minimised and suitable methods employed
- energy must be used efficiently
- effective measures must be taken to prevent environmental accidents.

A key requirement is that pig farmers are required to demonstrate that they are making use of Best Available Technique (B.A.T.) in respect of management practices as well as design and maintenance of buildings and equipment. Ethically the larger pig farm operators have a duty of care to the environment and its people. Monitoring environment amenity has a cost and those with the potential to pollute have to bear this cost. Large pig farms have to pay an IPPC initial application fee plus an annual subscription payment. There is speculation that, in time, these demanding E.U. rules will one day also apply to smaller pig units.

Pig farming's local impact

As a species *Homo sapiens* has a low tolerance threshold to pig unit odours. These smells are detected by people at low concentrations, long before higher concentrations of less offensive odours. Given this level of sensitivity there is a tendency to exaggerate the impact of pig odour on human health. Nevertheless, research has indicated that people living near to smelly pig farms were greatly troubled by odours associated with pig production. They suffered reduced vigour, increased fatigue, more confusion, more tension, depression and anger when compared to a control population (Schiffman *et al*, 1995). Environmental health officers in the UK consistently report that almost half of justified farm nuisance complaints are associated with pig production. Within the United Kingdom they have sweeping powers under Part III, Section 79 of The Environment Protection Act 1990. EHOs now have a statutory duty to inspect their areas to detect any statutory nuisances. They are also obliged to investigate all complaints of alleged nuisance. Ordinarily, assessment of smell has been mainly subjective, nevertheless, EHOs are empowered to serve an 'Abatement Notice' under Part III, Section 80 of The Environment Protection Act. This measure demands that the pig farmer brings about:

- the abatement of the nuisance
- the execution of such works or any steps necessary to put matters right.

It is significant that continued violation is regarded as a criminal rather than a civil offence and Magistrates' Courts have the authority to impose fines of up to £20,000 per offence against violators and offenders are also liable for consequential loss.

Is it fair that people should face criminal prosecution based on a subjective assessment from fallible human beings?

In recent years, laboratory assays have depended on panellists who grade smell in 'odour units' from a series of diluted samples of odorous substances tested against a clean air control. This assessment method cannot be used on site and it is time consuming and not "instant". Research is therefore underway to develop an instantaneous on-farm method of measuring pig slurry odours to help all those involved in pollution prevention.

Encouraging results on the measurement of odour have been obtained following a joint venture between The University of Manchester Institute of Science & Technology, North Wyke Research Station and Silsoe Research Institute (Persaud *et al*, 1996).

At the outset they identified the components responsible for unpleasant odours in normal pig slurry. Results are shown in Table 1.10.

Table 1.10 Substances mainly responsible for the smell in pig slurry

Chemical	Average concentration (µg/ml)
Acetic Acid	1365
Propanoic Acid	358
2-Methylpropanoic Acid	604
Butanoic Acid	237
3-Methylbutanoic Acid	301
Pentanoic Acid	90
Phenol	21
4-Methylphenol	62
Indole	6.6
3-Methylindole	3.7

When these smelly chemical compounds were then mixed together in the laboratory, it was found that the ph of the resultant solution was 3.1 compared to 8.2 in pig slurry. The alkalinity was therefore adjusted to this level using aqueous ammonia. Known concentrations of these volatile substances were then mixed with 20 polymers in a sealed vessel. Polymers were then used to adhere to these odorous substances and so facilitate their chemical identification.

It is hoped that this research will lead to the development and routine use of a portable monitoring device that would give 'real time' readings on a commercial pig unit.

Many consumers who enjoy wholesome pig meat would prefer not to have their amenity impaired because of nuisance or perceived nuisance from a neighbouring pig farm. Strict planning guidelines have for some time been in force which, in developed countries, increasingly have the look of 'catch-all' legislation. In the United Kingdom, Local Planning Authorities are required to abide by clearly defined policies set out in the document PPG7, ie. "Planning Policy Guidance 7, The Countryside – Environment Quality and Economic and Social Development". Local Planning Authorities (LPAs) use this detailed document to help formulate development plans. Planners have to balance the need to facilitate food production in the countryside without allowing significant impairment of the countryside and rural environment. Pig farmers are charged with efficiently and profitably producing pig meat without alienating those that purchase an end product which marketeers increasingly seek to distance from the farmyard.

Legally, new or extended pig buildings are regarded as "development" of which there are three categories:

- permitted development which can be carried out without informing the local planning authority (LPA).
- permitted development which requires prior notification to the LPA before building work starts.
- development which requires planning permission.

"Permitted development" arises in respect of minor extensions to existing buildings. New buildings are a different matter and invariably require planning permission and receive special attention if the intention is to site them within 400m (1300ft) of a dwelling or within 25m (85ft) of a classified road. Depending on the scale of operation and the particular locality, an Environmental Impact Assessment (E.I.A.) might have to be undertaken. An E.I.A. is used to inform decision makers and so help them make the best recommendations regarding agricultural development and its impact on the overall environment.

In particular, the establishment of large vertically integrated pig farms in America, especially in North Carolina, has attracted much attention from planners, politicians, concerned lobbyists and local residents. Increasingly, integrators wishing to establish large pig units try to identify potential sites with very low human population densities. Canada, the largest exporter of pig meat in the world, has also been the subject of planning constraints despite its broad acres. Nuisance issues arising from pig production are a serious impairment to marketing pig products. There are complex ethical issues regarding the responsibilities and respectful co-existence of those that strive to produce pig meat and those that choose to eat it.

Pig farmers therefore have to find a way forward that best meets the supermarket's specification for a particular brand of pig meat, avoids polluting the environment and enhances the welfare of the pig. Society which comprises many different cultures, writes the rule books which are regularly being modified, often to the dismay of those challenged with compliance.

In a review of the demands of modern livestock environmental design, Wathes *et al*. (2001) stated: *"It is almost as if the livestock farmer requires a licence to operate society's view on what constitutes acceptable agricultural practice which is changing rapidly in North America and Europe"*. In the same paper, the researchers draw attention to the fact that animals in their natural habitat make efficient use of time and energy to satisfy their various needs. There is an ethical aspect to the act of removing animals from an environment to which they have adapted via natural selection. How do natural foragers balance the new found time and energy at their disposal when feed is provided in a trough? Intensification reduces the uncertainty associated with providing the nutritional needs of the pig, both quantitatively and qualitatively. However, this lifestyle change burdens the pig with a vacuum of time which could result in unwelcome stereotypic behaviour. Ironically, ensuring that the pig is appropriately fed could actually result in farmers being accused of unethical practices. Such trains of thought are the prerogatives of affluent consumers and are generally not an issue for those short of good food. Furthermore, the pig did not evolve in an environment of high ammonia concentration, so, is it ethical to subject it to confinement in such an environment, particularly when the pig does not have the ability to recognise it as an alien environment? When pigs overheat in a natural environment, their behavioural adjustment often involves wallowing. Is it therefore ethical to impose an environment on a pig which cannot undertake this natural behaviour pattern? Could this shortcoming be justified in terms of the provision of a cleaner, healthier environment for the pig and increased wholesomeness for the consumer?

If the objective of large-scale profitable pig meat production at an affordable cost is accepted, is it unrealistic to demand that building designers should recreate a "natural" environment? The modern pig industry must face the challenge of providing an enriched artificial environment. This must be affordable, compatible with good pig welfare, acceptable to those working the system and able to satisfy the changing demands of sustainable agricultural production which is respectful of the global environment.

2

SOCIAL FACTORS

Pigs now found on commercial farms are the result of a long genetic journey started some 5000 years ago when the wild boar (*Sus scrofa*) was first domesticated. The main impact of people has been relatively recent. Some two hundred years ago, selective breeding aimed to maximise mature body size and fat production during a golden age of soap and candle-making and at a time when homes lacked thermal comfort.

The porcine pathway steepened somewhat in the second half of the twentieth century. There has been much emphasis on selection for leanness, efficiency of feed conversion, rapid growth rate and, more recently, eating quality.

Pigs are natural omnivores. They consume roots, grass, earthworms and other nutrients of plant and animal origin. Searching for food has a big influence on how pigs behave and where they live. Much of their time in the natural state would be spent rooting in the ground but there is also a degree of grazing and browsing.

Pigs like to graze and browse (JSR Genetics/BHR Communications)

44 Pig Environment Problems

The nasal disc of the sow is sufficiently rigid to withstand considerable resistance yet it is generously supplied with sensory receptors. In this way, it is well suited to rooting which it still undertakes, even when satiated. Anti-predator behaviour is generally well developed and this involves hiding and running. Having been originally supplied with gruesome tusks, pigs were well equipped to stand their ground and fight back if attacked. As a gregarious species, pigs choose to spend most of their lives in groups. Typically, in the absence of offspring, two to six pigs would form a stable social group in the wild.

The wooded and open areas inhabited by pigs of yesteryear are far removed from the environment needed to raise modern pigs for profitable meat production

Responsibilities of pig keepers

The availability of printed-circuits, "intelligent" environment monitoring systems and sophisticated electronic and electro-mechanical devices designed to optimise pig environment without human intervention brings both opportunities and challenges to the modern pig industry. It is understandable why stock-people sometimes feel side-lined and so choose to opt out of any on-going involvement in environment control. Whereas state of the art control systems can make a big impact on the precision of environment control, it is vital that those gifted with an empathy for pigs remain in overall control of their comforts.

People must assume responsibiltiy for pigs in their care (JSR Genetics/BHR Communications)

Silicon chips which pre-determine, e.g. temperature regimes, are only effective if correctly programmed and if they continue to function efficiently in changing environmental conditions within a dusty ammonical atmosphere which is notoriously hostile to sophisticated man-made devices. Those involved in the design, calibration, maintenance and routine use of automatic environment control systems must have a grasp of the biology of the pig. If stock-keepers are truly responsible for their pigs, they must endeavour to be smarter than the scientific equipment on which they are required to depend.

The fact that people must assume overall responsibility for the well being of pigs in their care is officially acknowledged. (DEFRA 2003)

Maintaining good environmental control involves an ability to mask vast changes in weather conditions and understand the complex anatomical and physiological changes that arise throughout the life of breeding and finishing pigs.

Challenges at birth

Within mammals, not only is body fat a source of energy, it also forms an effective insulation layer which helps to minimise body heat loss. A newborn infant starts life with a body containing around 15 per cent fat. Body fat content is particularly low in newborn pigs. Only about two per cent of the body of neonates comprises lipid (Table 2.1)

Table 2.1 Chemical composition of pigs (%)

	Birth	28 days	100 kg liveweight (ad-lib fed)
Water	77	66	60
Protein	18	16	15
Lipid	2	15	22
Ash	3	3	3

The table also indicates that by the time an *ad-lib* fed pig reaches 100 kg live weight, its percentage body fat is over tenfold the percentage at birth.

Anatomically and physiologically, slaughter pigs are markedly different to neonatal pigs and weaners. Hence the environmental needs of meat producing pigs change markedly between birth and slaughter.

The first milk or sow colostrum has a high protein content compared to conventional sow milk (Table 2.2). Within colostrum, large immunoglobulin protein molecules are present and these protect the pig against disease since piglets have a low immunity to disease at birth. The take-up mechanism involves direct absorption through the gut wall during the first 24 hours of life when the piglet consumes colostrum equating with around 20 per cent of its body weight. Immunoglobulins provide the neonate with passive immunity which protects it from maternal pathogens.

Table 2.2 Nutrient composition of sow colostrum and milk

	Colostrum (g/kg)	*Milk (g/kg)*
Water	700	800
Fat	70	90
Lactose	25	50
Protein	200	55
Ash	5	5

Body fat content is particularly low in newborn pigs (BPEX/Rattlerow Farms)

Another difficulty facing baby pigs at birth is that they do not enjoy the benefit of any blood-warming affectionate licking from their mother; this is unlike many other mammals. This oversight, combined with the piglet's lack of disease immunity, increases the likelihood of piglets becoming chilled at birth because of low air temperatures or high air speeds. Often, at this early stage of life, they become victims of disease, starvation or crushing by the dam. Providing the correct environment during this crucial time is, therefore, paramount.

Assuming the pig takes in colostrum and continues to suckle milk, by day 28 of life ordinarily body fat content will have increased sevenfold (Table 2.1). When piglets are between 21 and 28 days of age, immunoglobulin levels subside and the piglet assumes physiological responsibility for its own defence against disease and there is a change from passive to active immunity. This is a particularly challenging time for young pigs. Any shortcoming in the environment can be disastrous during this transitory stage, weaning is a further complication.

Whilst the sow has been lactating, the lipid : protein ratio in the piglet is built up to about 1 : 1, but after weaning this drops back to 0.5 : 1 as the piglet struggles to create an independent existence. Nature coped with these inherent short-comings by ensuring that sows were prolific and assumed a "survival of the fittest" approach. There are different priorities in commercial farming. Commercial pig production is about maximising productivity by providing the optimum environment, pampering pigs when they are vulnerable and making them feel comfortable when conditions favour growth.

The trauma of weaning

At weaning, piglets lose the benefit of mother's milk sympathetically presented within a comforting social and climatic environment. Prior to weaning the temperature in the creep area should be reduced, this helps stimulate intake of solid feed and prepares the pigs for the withdrawal of milk. Weaning is a traumatic time for piglets and there is inevitably a check in nutrient intake. Careful management of the post-weaning environment must help ensure that this setback is minimised.

Maintenance, growth and body composition

Just as covering fixed costs are the first priority in any business, there is an "up-front" cost of simply maintaining the pig. The concept is shown in Figures 2.1 and 2.2.

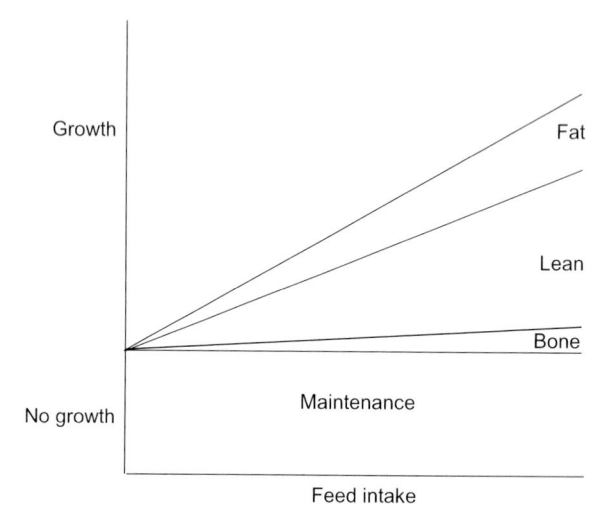

Figure 2.1 Maintenance and growth in relation to nutrition.

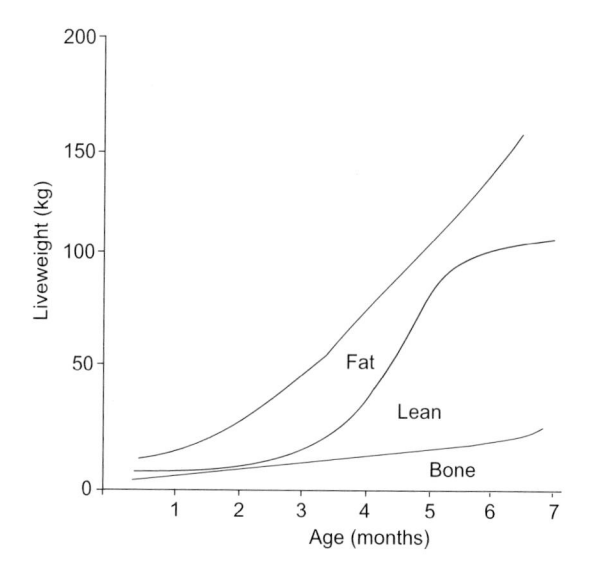

Figure 2.2 Changes in carcass composition over time.

Initially nutrients are converted into bone and lean but, as pigs get older and consume more food, fat is deposited within the carcass.

Genetics, health status and nutrition have a big bearing on the relative deposition rates of lean and fat and the resultant composition of the carcass. If pigs are generously fed, fat deposition increases (Figure 2.3)

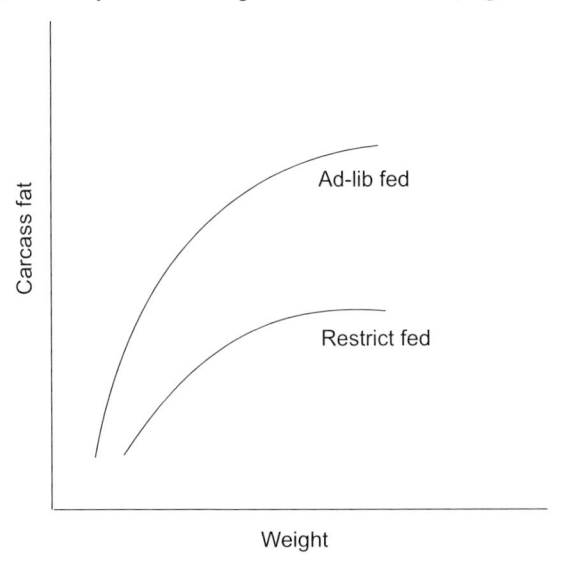

Figure 2.3 Body fat in relatin to feeding regime.

Environmental factors such as provision of a dry bed, lack of draughts and temperature at pig level, together have a big influence on feed intake and carcass composition.

Feed conversion efficiency and daily liveweight gain

Figure 2.4 provides a further insight into how pigs grow and the efficiency of feed conversion. At low feed intakes there is little daily liveweight gain since feed is being used to maintain the pig, hence feed conversion efficiency (FCE) is poor. As feed intake increases, more nutrients become available to grow the pig so there is an associated increase in daily liveweight gain and an improvement in feed conversion efficiency. This situation progresses until the process of fat deposition takes off. Modern genetics aim to maximise the production of lean and delay the onset of fat deposition. Depositing fat mops up energy, but eventually a stage arrives whereby the amount of energy demanded exceeds that linked to the earlier growth. The result is a deterioration in feed conversion efficiency. In commercial situations the appropriateness of the pig environment has a big influence on the theoretical concepts outlined in Figure 2.3. Shortfalls in the pig environment make nonsense of this stylised illustration.

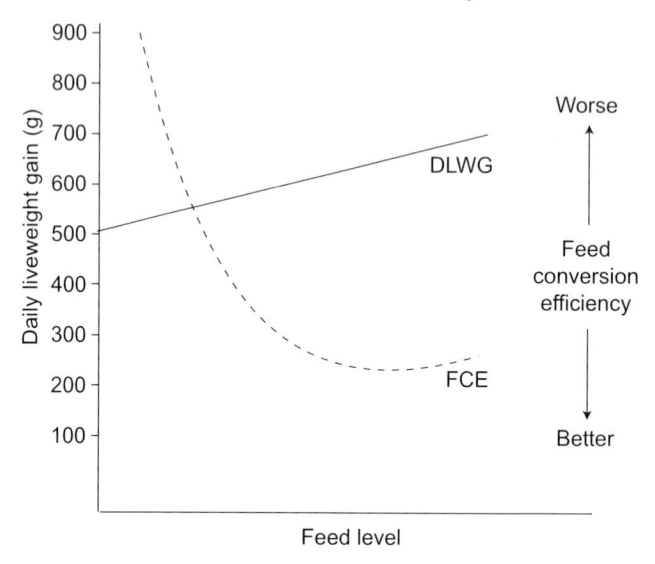

Figure 2.4 Influence of feed level on feed conversion and growth rate.

The changing needs of breeding stock

The lifespan of a breeding female pig is around eightfold that of a finishing pig. Hence, pigs destined for breeding must be built to last and throughout the breeding cycle appropriate environmental conditions must be provided for them. Figure 2.5 provides a summary of life's main events for a breeding sow.

Modern gilts are late maturing and their lifetime productivity is more likely to be enhanced if fat reserves are correctly established and conserved whilst avoiding the depletion of lean.

Ideally gilts should have a 16-18mm P_2 backfat thickness at service and weigh no less than 130kg liveweight. Thereafter their nutritional management must be specific but the common aim is for backfat conservation, the avoidance of excessive weight gain in pregnancy and the avoidance of backfat and weight loss during lactation. Table 2.3 lists some simplified guidelines for sow feeding throughout the breeding cycle.

Social Factors 51

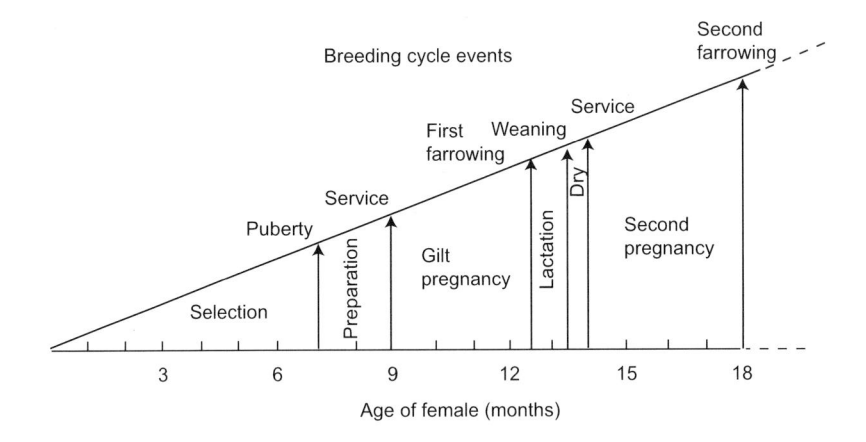

Figure 2.5 Key events in the life of a breeding sow.

Table 2.3 Simplified guidelines for sow feeding

➢ Never let a sow become thin

➢ Regularly assess body condition

➢ Monitor the environment

➢ Try to think one step ahead in the breeding cycle

➢ Feed 'low' levels during early pregnancy

➢ Adjust feed levels +/- during mid-pregnancy

➢ Increase feed levels from day 100

➢ Reduce feed levels before farrowing

➢ Increase feed allocation gradually in lactation

➢ Continue to increase this allowance as long as the sow will eat

➢ Feed *ad-lib* from weaning to service

The correct feeding of sows is becoming a very finely tuned science and its efficiency is increasingly dependent upon the provision of the optimum pig environment at the various stages in the breeding cycle.

Recognising normal and abnormal behaviour

The intensification of pig farming has increased the likelihood of disease incidence on account of the concentration of viruses and bacteria within confined air spaces. Similarly, the reduction in floor area associated with intensification has had an impact on the behaviour pattern of modern pigs. An understanding of the behavioural traits of pigs in their natural surroundings and an appreciation of what is deemed to be an acceptable degree of modification of this behaviour to cope with the associated intensification helps the observer to define and recognise abnormal behaviour patterns. Having recognised abnormal behaviour, the pig keeper should then identify the cause of the aberration and strive to remedy any shortcomings. The need for political correctness has eroded the impact associated with that emotive, unique word "stockmanship" which seems to bring together all that is good about the relationship between people and animals entrusted to their care.

Few pig keepers have had the opportunity of observing normal behaviour in a natural free-range situation. Nevertheless, they must be able to reassure themselves by recognising acceptable general behaviour of pigs, how they interact socially, how they compete for feed, deal with their sexuality and respond to new situations which might be stimulating, boring or intimidating. Stressful situations, the effects of medication, nutritional extremes and environmental influences impact on the physiology of the pig in addition to its behaviour patterns.

Space has a massive impact on pig behaviour. In a free-range environment, pigs make use of available space to seek food. Under natural circumstances pigs spend 30-40% of their time foraging by grazing and rooting (Peterson. 1994). The provision of food lessens the need for foraging; farmed pigs consequently only spend around a third of their time foraging compared to those in their natural environment (Tynes, 1997). This creates a time vacuum and the pig attempts to satisfy its natural curiosity by taking an increased interest in the fabric of its artificial environment. Chewing anything or chewing nothing become a favourite pastime, as does the shaking and often damaging or total destruction of anything accessible to the pig.

Provision of food creates a time vacuum (Colin Woolridge)

Space is also a key factor in the dunging behaviour of pigs. In the free-range situation, pigs had ample opportunity to find solitude in order to defaecate with minimum risk from aggressors. Intensification has diminished this facility. However in an indoor situation, given adequate space, pigs tend to organise themselves so that they use a "safe" dunging area away from their food, water and sleeping areas (Baldwin, 1969). A practical difficulty arises, however, in that pigs often drink, defaecate and urinate in close sequence and this can result in dunging in wet, dirty areas near water bowls and nipple drinkers. Pigs enjoy comfort. Hence they choose to lie in an environment deemed to be comfortable by themselves. Having selected that environment, pigs then elect to dung "somewhere else", providing the space is available. During periods of heat stress, pigs feel more comfortable in wet areas which, despite these often being dirty, are still preferred as a temporary lying area. Extremes of stocking density and the existence of stress have a marked influence on the excretory behaviour of pigs (Harker *et al.*, 1994).

Overstocking can lead to:

- Greater variation in pen weights

54   *Pig Environment Problems*

- Vices
- Physical injuries/leg weakness
- Dunging in the lying area
- Lying in the dunging area
- Reduced growth rates
- Poorer feed conversion efficiency
- Rectal prolapses
- Respiratory and other diseases
- Dehydration
- Hyperthermia
- Suffocation
- Condemnations

Cold pigs, poor feed conversion, poor growth rates and dirty lying areas are often associated with under-stocking.

Pig behaviour can also be massively influenced by the type of feeding system employed. Floor feeding tends to lead to faecal contamination of the skin which is a pre-disposing factor for salmonella infection. Pig keepers employing floor-feeding systems tend to develop feed allocation routines which are more concerned with keeping floors and pigs clean, rather than maximising feed intake.

The impact of pen layout

Siting of ad-lib feed hoppers and drinkers has a big impact on pig behaviour. Figure 2.6 outlines four different locations for an ad-lib hopper.

The layout in Pen A is typical of a set-up with manual feeding of growers or finishers where the facility to top up hoppers from a feed passage dominates the layout. A few minutes observation of such a layout would suggest that pigs seem to be in perpetual motion since the busy track-way to the feed hopper traverses the lying area. Layout B normally results in less traffic across the lying area but often still more than a tolerable amount of disturbance, even though most of it is confined to one end of the pen. Layout C can be a disaster if dung channel width is restricted. Pigs tend to overcome the problem of eating within a cluttered dunging area by dunging behind the feed hopper but within the lying area. Layout D generally works well when dung passage width is at least 2.4m (8ft) and the hopper is set on a plinth.

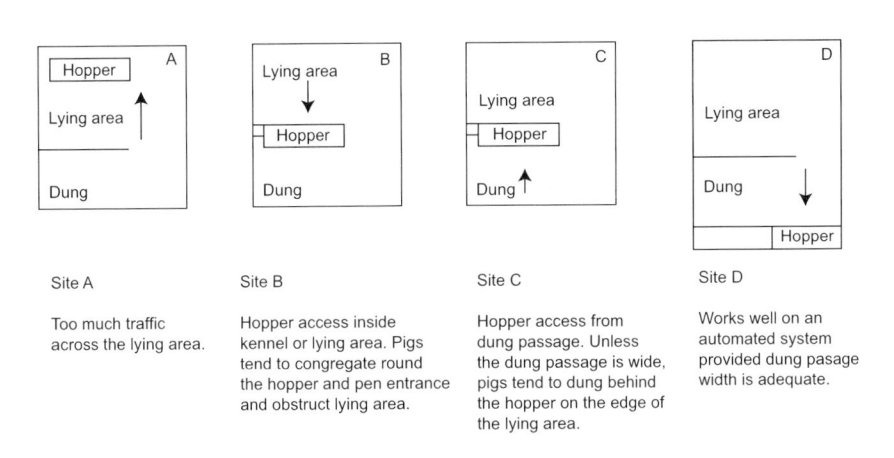

Site A	Site B	Site C	Site D
Too much traffic across the lying area.	Hopper access inside kennel or lying area. Pigs tend to congregate round the hopper and pen entrance and obstruct lying area.	Hopper access from dung passage. Unless the dung passage is wide, pigs tend to dung behind the hopper on the edge of the lying area.	Works well on an automated system provided dung passage width is adequate.

Figure 2.6 Alternative position of feed hoppers.

Siting of *ad-lib* hoppers has a big impact on pig behaviour (Colin Woolridge).

56   *Pig Environment Problems*

Comfort and confrontation avoidance is important to most pigs. No doubt the work of Phillips and Fraser (1987) undertaken in Canada has had an influence on the Dutch "Varkensbesluit" national regulation about minimum requirements for pig housing in response to Council Directive 91/630/EEC (Dutch Meat Board 2003). The Canadian researchers investigated the possibility of a split-level pen for growing and finishing pigs. By providing a ramp to a higher level, pigs were given a choice of thermal environments, different floor surfaces, a more relaxed stocking density and an increased opportunity for exercise.

An existing 2.6m by 4.9m (8ft 6in x 16ft 1in) totally slatted pen was adapted to take an additional raised floor. The layout is shown in Figure 2.7.

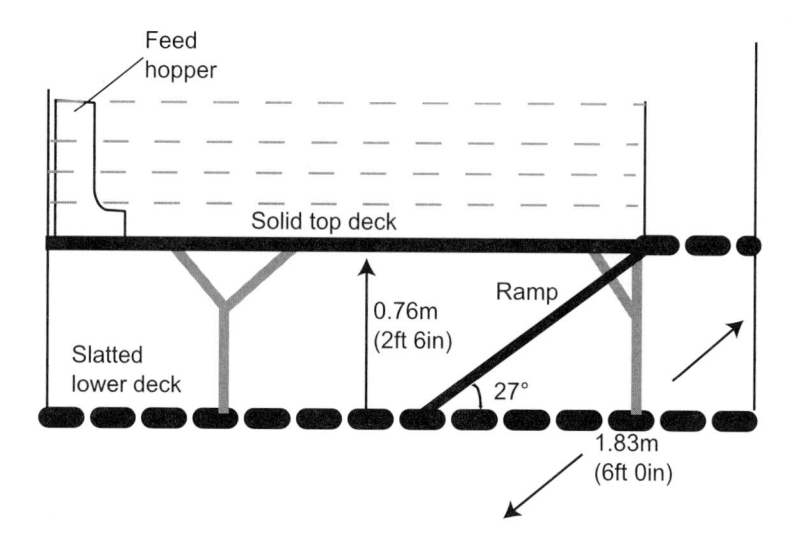

Figure 2.7 Section through free-access split-level pen.

The lower level floor had concrete slats and provided three nipple drinkers. Apart from a small slatted "doormat" area at the top of the ramp, the upper level had a solid floor and feed hopper plus an extra nipple drinker provided over the raised slats.

The raised floor provided an additional 9m² (97ft²) and when the pen was stocked with 24 finishing pigs, the total floor area per pig was a generous 0.88m² (9.5ft²). Results of the trial are shown in Table 2.4.

Table 2.4 Pig performance in different pen layouts

	Split Level	Control
Days to slaughter	153.3	154.6
Starting weight (kg)	25.6 +/- 4.35	24.4 +/- 3.01
Slaughter weight (kg)	99.2 +/- 5.74	96.7 +/- 5.84
Daily liveweight gain (g/day)	878 +/- 82.2	849 +/- 94.4
0-7 days	5.46 +/- 1.56	5.09 +/- 0.84
Feed conversion efficiency	2.70 +/- 0.095	2.72 +/- 0.054
Carcass index	100.7 +/- 3.82	100.6 +/- 3.11

Pigs readily adapted to the two-tier system and, when compared to ad-lib fed pigs in a conventional pen, differences in live-weight gains during the first week of the trials were not significant. Results summarising pig behaviour are shown in Table 2.5 and indicate that the pigs spent about as much time overall at the two levels.

Table 2.5 Pig behaviour in split level pen

| | | Time % | |
Level	Active	Resting	Total
Upper	12.6 (10.4-14.3)	31.4 (12.2-463)	44
Lower	9.4 (5.5-14.0)	46.6 (37.8-59.4)	56
Total	22.0 (15.9-28.3)	78.0 (71.0-84.4)	100

However, pigs elected to spend more resting time in the lower level, no doubt trying to avoid "traffic" in the upper level where pigs were feeding. Relative comfort on slats was preferable to constant disturbance on a solid floor. No doubt, had the hopper been relocated in the slatted lower deck, the more tranquil upper deck would have become the preferred lying area.

Social factors

Pigs are social animals and generally seek the company of other pigs. When they are isolated from the rest of the group they become agitated (Fraser, 1974). Significantly, Section 2 (27) of UK Pig Welfare Code states: *"Boar pens shall be sited and constructed so as to allow the boar to turn round and to hear, see and smell other pigs, and shall contain clean resting areas"* (DEFRA 2003). Outdoor pig keepers observe that the weaner pig, in particular when unexpectedly isolated on the wrong side of a barrier, tends to panic and will go to great lengths to re-join its peers.

In common with many domestic species, pigs have a social dominance hierarchy and this ordered behaviour tends to decrease agitation and reduces aggressive encounters but does not eliminate them. The social hierarchy arises as a result of agonistic interactions. Typically there are about 12 pigs in each social group but hierarchies with up to 18 pigs have been observed (Ewbank, 1969). The structure of the social hierarchy is generally linear but not always. Sometimes "equals" can be identified within a group and, on occasion, dominance circles or "ganging-up" can arise (Ewbank, 1969). Dominance has its rewards. Dominant pigs take control of the most comfortable lying areas which tend to be the most distant from the dunging areas. These privileged pigs tend to be first in the queue when they choose to feed (Brouns & Edwards, 1994).

When large numbers of pigs are kept within the same pen, small "family" groups tend to be established and pigs within the respective groups choose not to interact with those in other groups (Ewbank, 1969). The advent of dry sow housing systems involving large groups of sows retained within a vast yard and fed from an electronic feeder station provides pig keepers with a useful opportunity to verify these research findings.

Recognition involves sight and smell and pheromones are thought to play a key role. Ewbank and Meese (1971) found that dominant pigs could be removed from a group for up to three weeks and then be re-introduced with a minimum incidence of confrontations. Shy pigs, if removed for as little as three days, are bullied and often severely attacked upon their return to the pen. Apart from the risk of infection spread, this research observation makes nonsense of the practice of hospitalising sick pigs and then re-introducing them to their original groups.

The Standing Committee of the European Convention for the Protection of Animals kept for Farming Purposes has flagged-up concerns about aggression arising when pigs are mixed and suggests that: "Visual barriers within group pens which allow pigs to hide from aggressors should be provided at least during mixing".

When pigs are kept in large groups, small 'family' groups tend to be established (Colin Woolridge).

The Dutch "Varkensbesluit" national regulations also highlight the need for stable groups:-

"Weaners and rearing pigs shall be kept in stable groups. The aim of stable groups is to create a stable social environment as soon as possible. Pig welfare benefits from a predictable social environment with a stable social hierarchy".

Boredom is the scourge of modern pig keeping and is likely to be reduced if the pig environment is enriched. Barren, fully slatted pens impose a behaviour pattern far removed from free-range situations. In more natural environments, pigs eagerly explore territory in an effort to locate an environment rich with forage and capable of providing comfortable shelter.

The Standing Committee of the European Convention for the Protection of Animals has addressed the issue of environment enrichment. The Committee suggest that all pigs should have access at all times to adequate amounts of materials for investigation and manipulation, including rooting.

Materials listed include straw, hay, maize chaff, grass, peat, earth, wood and bark. The Standing Committee viewpoint gives an indication on current European opinion regarding the need for provision of manipulable materials. This consensus could well be used as a basis for changes in European law when the EC review the current EU pig welfare regulations in 2005.

Vocalisation

As a species, pigs are very vocal, particularly when kept in confined spaces. Vocalisation is used specifically to locate other members of the herd and as a more general method of social communication. Types and intensity of vocalisation vary, particularly between lactating sows and their litters. Behaviourists have studied the different sounds made by pigs and have related them to specific responses within their environment, particularly when changes are underway or anticipated. Some pig noises are gender specific. Boars have a typical "courting song" (Baldwin, 1969) and farrowing house attendants are very familiar with the sow's nursing sounds (McBride, 1963). Pig vocalisations can be subdivided into:

- Feeding,
- Warning,
- Location, and
- Alarm.

Table 2.6 provides a summary of research on the different types of vocalisation. A characteristic of the species is that vocal responses are markedly different in pigs kept in groups compared to those separated from their peers. Enlightened modern pig keepers strive to interpret the various pig vocalisations and use them as a means of establishing and maintaining stable groups within a pig friendly environment. Looking to the future, the prospect of intelligent software being developed which will log noises made by housed pigs, and so inform the pig producer of changes in behaviour, seems increasingly likely.

The impact of temperature

Wild pigs modify their behaviour patterns in response to changes in

Social Factors 61

Table 2.6 Summary of pig vocalisations after Grauvogl (1958) and Kiley (1972)

Classification	Situation	Description	Occasion
Feeding	Calling	Sows grunt fast and rhythmically	Beckoning litter to suckle (sometimes also warning the litter)
	Milk let down	• Fairly rapid grunting • Same pitch as calling sound but less regular • Sharper grunting	• Early stages of udder massage • Milk flow • After udder massage
	Hunger	Adult makes a bright uniform cry Piglet makes either high pitched squealing or low growling	
Warning	Signalling possible danger	A single short bark common to all pigs	Arises after pigs are startled. Pigs in immediate group 'freeze' until the perceived threat abates.
	Anger	Adults make sharp 'quacking' sounds often in unison. Piglets make short 'barking' sounds	Uncomfortable sensation. Accompaniment to fighting with other pigs
Alarm	Pain or fear	Voluminous sound rising to a piercing cry in adults, a high pitched sound in suckling pigs and weaners or a low 'growling'.	Rough handling or response to being driven or restrained. Separation of piglet from the litter or 'family group'
	Indignation or pain	Low rolling growl in sows	Acknowledgement of the need for self-defence, especially in pregnant sows
	Defensive	Loud rapid grunting from open jaws in the adult	During the act of self-defence or defence of offspring

62 Pig Environment Problems

Table 2.6 Contd.

Classification	Situation	Description	Occasion
Location	Contentment	Individual low grunting	An expression of approval especially during back scratching
	A prelude to social contact	Individual low, irregular grunts in sows and boars and a 'soft snore' in piglets	Pinpointing whereabouts of group mates
	Interest	High, sharp grunts	Arousal of interest

ambient temperature, and in situations of extreme heat, modify their feeding times by choosing to feed at night rather than during daytime (Baldwin, 1969). This behavioural response arises because pigs have only one row of sweat glands located above their snouts and these glands are non-thermoregulatory (Meese *et al.*, 1975). In an outdoor situation when pigs feel too hot, generally they initially seek shade.

Outdoor pigs seek shelter when they are too hot (Galebreaker Products).

If this does not lower their body temperature sufficiently, then wallowing in mud takes place. Pigs kept indoors or in confined spaces do not have this choice. Consistent with the maxim "pigs lie where they are comfortable and dung somewhere else", under testing circumstances, behaviour patterns are adjusted. Drinking water is deliberately spilled on the pen floor and pigs choose to lie in wet areas and so lose body heat by evaporative heat loss (Baldwin, 1969, Tynes 1997). Huddling is a behavioural response to low temperatures and is also influenced by floor surface, availability of bedding and air speed at pig level. In hot weather, a typical behaviour pattern is for pigs to maximise their surface area by stretching. Boon (1981) concluded that the lying space requirement for growing pigs increased by up to 15 per cent as air temperature increased by 6°C above the Lower Critical Temperature (LCT) when other environmental factors remained constant. This presents a practical problem for the commercial pig farmer. Ideally the number of pigs housed in each pen in warmer weather should, on average, be lower than that housed during cold periods.

This plays havoc with throughput calculations and delivery schedules. When lying areas are bedded, sometimes pig keepers reduce the amount of bedding or totally remove it during hot weather. This allows pigs to dissipate heat directly through, e.g. a concrete floor surface. In practice it is difficult to anticipate likely increases in ambient temperature and there is a danger that bedding will not be replenished when temperatures fall to more normal levels.

The impact of feed

Pigs modify their feeding behaviour in response to their individual needs for energy and amino acids. Given the opportunity, ie, the provision of ad-lib feed, pigs tend to eat to satisfy their overall energy needs irrespective of the digestible energy content of the food (Cole and Chadd, 1989). However, in younger pigs, gut capacity is usually a limiting factor.

When pigs are provided with ad-lib feed, they tend to satisfy their overall energy needs (Simco Systems).

Feeding behaviour can also be modified by the introduction of various flavouring compounds to pig rations. Particularly during the post-weaning period, when lack of feed intake can be a problem, flavouring agents are often used to stimulate appetite. In particular, "cheesy", "meaty" and "sweet" flavourings are known to boost feed intake during this period (McLoughlin *et al.*, 1983). Older pigs prefer sucrose to glucose and lactose solutions (Kare *et al.*, 1965). If sucrose is added to water, the intake of water has been known to quadruple within hours. The pig's affinity for sucrose-flavoured water far exceeds nutritional demands and the physiological need for water (Kennedy and Baldwin, 1972).

The social nature of the pig has an influence on its feeding behaviour and this must be taken account of when pig housing and feeding systems are designed. When a pig is isolated in a pen and provided with a generous allocation of feed, even though apparently nutritionally satisfied, the subsequent introduction of a still hungry pig results in the original pig ingesting more feed (Hsia and Wood-Gush, 1984; Bigelow and Houpt, 1988). The same behaviour pattern has been observed when pigs are housed in adjacent pens (Gonyou *et al.*, 1992).

Ad-lib feeding systems provide an opportunity for low-ranking pigs to consume sufficient feed to meet their nutritional requirements (Brouns and Edwards, 1994). When feed is restricted, competition is increased. Hence, with once and twice daily feeding regimes, low ranking pigs are disadvantaged.

Dominant pigs assert themselves at feeding time. If feed is rationed, they eat more than low-ranking pigs and grow faster (Brouns and Edwards, 1994; Hsia and Wood-Gush, 1984). However, factors other than social dominance are also involved. Dominant pigs tend to be heavier and heavier pigs have a higher maintenance requirement and so tend to eat more (Bigelow and Houpt, 1988).

Stereotypic behaviour of pigs

If the welfare of pigs is compromised, stereotypical behaviour often results. Stereotypes have been defined as "behaviours that are relatively invariant, regularly repeated and without obvious function" (Kiley-Worthington,

1977). Such aberrant behaviour often arises because of boredom, frustration, arousal or social isolation (Dantzer, 1986; Lawrence and Terlouw, 1993; Wemelsfelder, 1990). Frequently, stereotypes involve oral manipulations such as vacuum chewing, nosing or chewing pen-mates or any fixture or fitting within the pen.

Intensification of the pig industry is linked with an increase in stereotypes, yet outdoor sows regularly indulge themselves in stone chewing. In particular, feed restriction results in pigs spending more time standing, they are more active, they root over the floor covering and sometimes attempt to destroy the floor surface (Day *et al.*, 1995). Increased oral activity seems to occur because pigs are seeking food and it often continues after feeding time as the pigs seek more food (Rushen, 1984). Excessive drinking is a form of stereotypic behaviour and this can be reduced if pigs get a feeling of satiety after being fed straw or other fibrous food. Pigs adopt "coping mechanisms" which are physiological and behavioural patterns in response to stress.

The incidence of vices

Tail biting has pre-occupied pig farmers and scientists for years, its incidence is somewhat variable (Van Putten, 1969; Petersen *et al.*, 1995). If tails are docked or tipped, pigs are less likely to tail bite. Non-stimulating environments, boredom, lack of bedding, inappropriate temperature, excess stocking density, poor ventilation or sickness have all been linked with tail biting (Van Putten, 1969; Petersen *et al.*, 1995). Injuries arising from tail biting can become infected and result in the condemnation of the whole carcase at slaughter (Van Putten, 1969). In some incidences of tail biting, pigs are thought to be motivated by the taste of blood, often this is followed by cannibalism and it is commonplace for a tail bitten pig to be savaged by pen-mates.

On some occasions, dietary imbalance has been blamed for tail biting outbreaks. A lack of protein or lack of fibre in the diet has been implicated. Unappetising formulations could result in pigs developing "ration boredom", causing them to crave for alternative flavours and dried blood could fulfil their need (Fraser, 1987a; Fraser, 1987b). Experimentally it has demonstrated that when pigs were fed salt-depleted rations, their interest in dried blood increased (Fraser, 1987b). On some commercial farms tail biting has been eliminated by increasing the salt content of the ration. Research undertaken on rats, rabbits and sheep has indicated elevated levels of adrenocorticotrophic hormone which is normally associated with stress. If the same mechanism operates in pigs, it could well be that any stressful situation will increase the likelihood of an outbreak of tail biting. The tip of a pig's tail is not well supplied with nerves. Hence, if a pig nips the insensitive end of the tail of a pen mate, the need for a defensive response or even a counter-attack is diminished. If left unchecked. the aggressor further indulges itself.

Tail-docking and tail-biting are best avoided. However, when tail docking proves necessary, the procedure should be undertaken within the first week of life. It should be regarded as a last resort measure, the consequence of providing a sub-optimum environment. Every pig keeper should have a strategy in place for coping with tail biting. In the event of an outbreak, the following questions need answering:

- Is the problem widespread or localised. just involving certain pigs in a particular pen?
- Which pens or buildings are involved in the current outbreak?
- Do records indicate that the problem has arisen previously in the same place, or is this a new experience?
- Is water supply adequate and has it been disrupted recently?
- Is the environment too barren and in need of enrichment?
- Are pigs subjected to excessive temperature fluctuations?
- Are they lying in a draught?
- Is the pen or building overstocked?
- Is there adequate ad-lib hopper space?
- If pigs are restrict-fed, is there excessive competition at feeding time?
- If automatic feeding is installed, is it working properly?
- Is there too much light in the building and could this be upsetting the pigs?
- Are dust and gas levels unacceptably high?

In the event of an outbreak of tail biting, the early identification of the instigator is vital. Aggressors should be removed from the pen and isolated. Pigs with bitten tails should be removed from mainstream production and be housed in hospital pens where they must remain.

Sexual behaviour and parenting skills

Before a sow comes into oestrus, besides well defined physical changes, there are also marked behavioural changes:

- The female becomes agitated and easily disturbed
- Other sows try to ride the pro-oestrus female but she refuses to stand
- There is increased vocalisation

Typical behavioural characteristics of an oestrus sow include:

- A general increase in activity
- The female starts to mount other sows and, if mounted herself, seems more willing to stand
- A high pitched grunt is heard
- The appetite diminishes
- The female stands firmly to back pressure, particularly if a boar is present
- Ears are pricked in Large White (Yorkshire) sows
- Rub marks are often observed on the skin
- The female is keen to rub against stockpersons
- The female allows coitus with the boar

Boars exhibit courting behaviour once a sow in oestrus has been identified. Typical behaviour of the boar involves:

- The boar frequently urinates in the presence of the sow
- He chews and produces a foamy saliva
- The boar rests his head on the back or shoulder of the sow
- The boar gently bites the ear of the sow and, less gently, lifts the sow by placing his snout between the sow's legs

The behaviour of sows around parturition has a big influence on the environmental requirements of farrowing and nursing accommodation. The sow's behaviour pattern changes markedly in a matter of days:

- **Before farrowing**:- During a period of 24-16 hours prior to farrowing the sow chooses to isolated herself and select a nesting site.

- **Nest Building**:- Around 16 hours before farrowing, the sow engages herself in vigorous nest construction activity which generally ceases about 3 hours before farrowing.

- **Parturition**:- During a 3 hour period before farrowing until the end of farrowing, the sow is totally pre-occupied with producing her litter.

- **Nest Occupation**:- The time from farrowing until 10 days after is a time of bonding, the sow and her litter choose to keep together in close proximity to the nest.

- **Social integration**:- Thereafter, given the opportunity, sows will increasingly spend time away from the litter and piglets will attempt to socialise with other litters.

Necessarily, the design of farrowing pens is very much influenced by the behaviour of the sow and litter at this vital time. Inevitably, there are conflicting requirements. Should the sow be denied the opportunity of nest building by housing her on a non-bedded slotted floor? It is generally acknowledged that this eases the workload, improves hygiene and piglet health, but the sight of a sow trying to make a bed on a slotted floor is repugnant to many animal lovers. In a natural situation the sow attaches great importance to hygiene by depositing faeces several metres from the nest. Given the fact that after ten days, sows choose to spend time away from the nest, is it fair to keep the dam fully crated until weaning? Whilst the argument in favour of preventing the crushing of piglets is clear-cut, other issues remain open to discussion. The key issue is that stockpersons should have a grasp of normal pig behaviour from birth to slaughter, understand the pig's need to express sexuality and endeavour to provide and maintain the optimum environment as the social demands of the pig change.

3

THE IMPORTANCE OF THERMAL COMFORT, LIGHT AND ELECTRICAL SAFETY

Understanding pig physiology

Pigs are homeotherms, i.e. warm-blooded animals and provided the environment avoids extremes of heat, cold and relative humidity, they have a good biological control of body temperature. Pigs strive to maintain a deep body core temperature over a range of environmental conditions. The concept is shown in Figure 3.1.

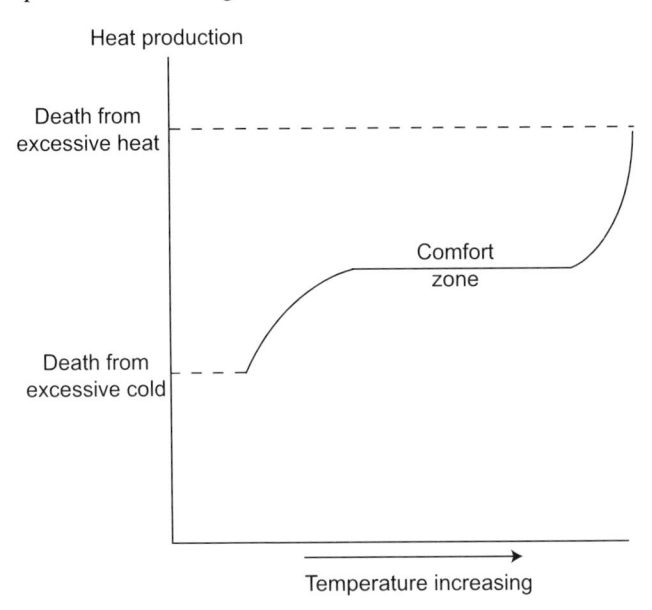

Figure 3.1 Heat production and body core temperature related to ambient temperature.

This relatively wide temperature range is known as 'the zone of thermal neutrality' and it lies between the Lower Critical Temperature (LCT) and Upper Critical Temperature (UCT), (Figure 3.2). Hence, baby pigs are particularly sensitive to the cold since they have a high LCT.

72 Pig Environment Problems

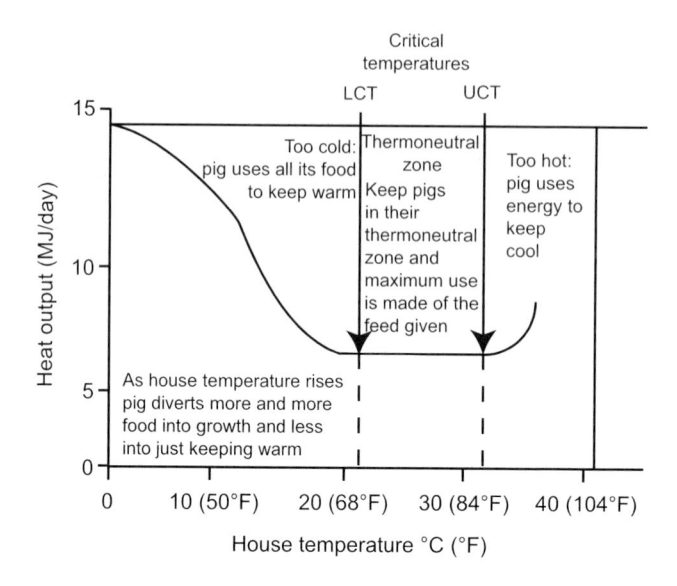

Figure 3.2 The principles of critical temperature.

[Graph: Heat output (MJ/day) vs House temperature °C (°F). Shows critical temperatures LCT and UCT bounding the Thermoneutral zone. Labels: "Too cold: pig uses all its food to keep warm"; "Thermoneutral zone — Keep pigs in their thermoneutral zone and maximum use is made of the feed given"; "Too hot: pig uses energy to keep cool"; "As house temperature rises pig diverts more and more food into growth and less into just keeping warm". X-axis: 0, 10 (50°F), 20 (68°F), 30 (84°F), 40 (104°F). Y-axis: 0, 5, 10, 15.]

Conversely, pigs susceptible to heat stress have a low UCT, i.e. they suffer from excess heat at an earlier stage than more heat tolerant pigs.

If a pig becomes chilled, food and body tissue are broken down in an attempt to maintain the core temperature. Whilst this is unproductive commercially, it has its biological limits as well. If severe chilling persists, biochemical processes within the pig ultimately prove incapable of even achieving the lower reaches of the comfort zone and eventually death results from hypothermia. When pigs get too hot, there is only a very limited facility for evaporative heat loss by sweating. That's the reason why pigs wallow. When water evaporates from a surface, the energy-charged hottest molecules are lost first – which is the very reason why dogs pant. Unless there are environmental provisions to cool down a severely heat-stressed pig, there is a strong possibility that the pig will expire from hyperthermy.

There is a close relationship between feed intake and critical temperatures and in particular much interplay between LCT and nutrition at weaning. The latter is often regarded as a complex physiological process involving feed, temperature and social interactions. Prior to weaning pigs generally enjoy a comfortable social and climatic environment. Sow milk and a

Thermal Comfort, Light and Electrical Safety 73

highly digestible energy rich creep feed are usually on offer in a warm environment and plenty of fresh water is available. This comfortable status is disrupted at weaning and there is a sudden elevation of the Lower Critical Temperature as depicted by the inflection on the graph shown in Figure 3.3.

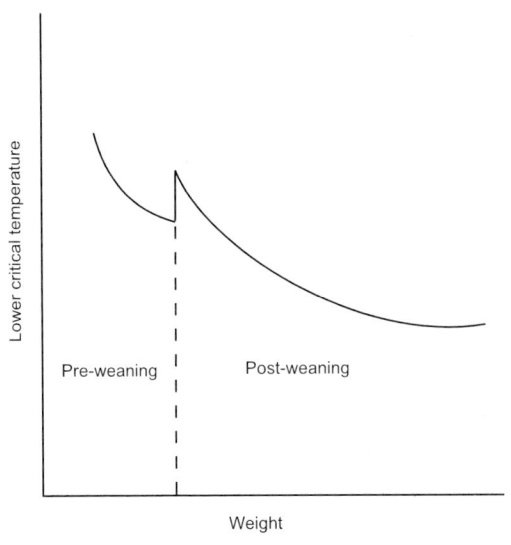

Figure 3.3 The impact of weaning and weight gain on lower critical temperature.

Highly digestible creep feed helps warm pigs before weaning (Tuckbox/BHR Communications).

74 Pig Environment Problems

This arises because creep feed and sow's milk are withdrawn at weaning and the weaners face the challenge of establishing themselves within a new social group. Figure 3.4 is based on research undertaken at Scotland's Centre for Rural Building.

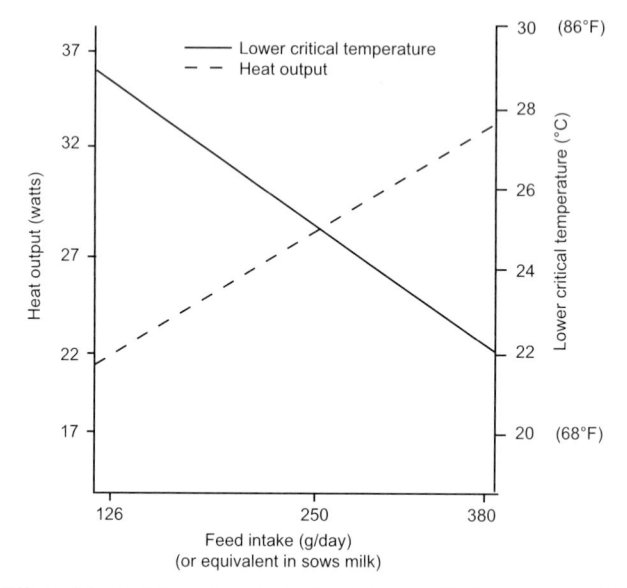

Figure 3.4 Effect of feed intake and heat output and lower critical temperature (LCT) on a 5kg pig.

The broken line on the graph illustrates that a lightweight poor competitor 5kg (11 lb) pig eating only 126g (4.4 oz) per day reduces its body heat production to just 22 watts in an attempt to keep warm. At such a low feed intake it assumes a LCT of around 29°C. If it were physiologically possible for the same weight of pig to ingest 380g (13.4 oz) per day of starter feed, without incurring digestive upset, the pig would be comfortable at a much lower temperature. Heat output would increase to 33 watts and in sympathy the LCT would fall to 22°C, i.e. increased feed intake would reduce the Lower Critical Temperature.

The impact of flooring

Since pigs at rest invariably have a degree of contact with a floor surface, the nature of that surface will have a big influence on how thermally comfortable the pig feels and how much heat is lost to that floor surface.

Thermal Comfort, Light and Electrical Safety 75

In cold conditions, a dry straw bed has an insulation function and this helps prevent heat loss through the underlying surface. Hence a straw-bedded pig has a reduced LCT compared to a pig on a perforated floor (Figure 3.5)

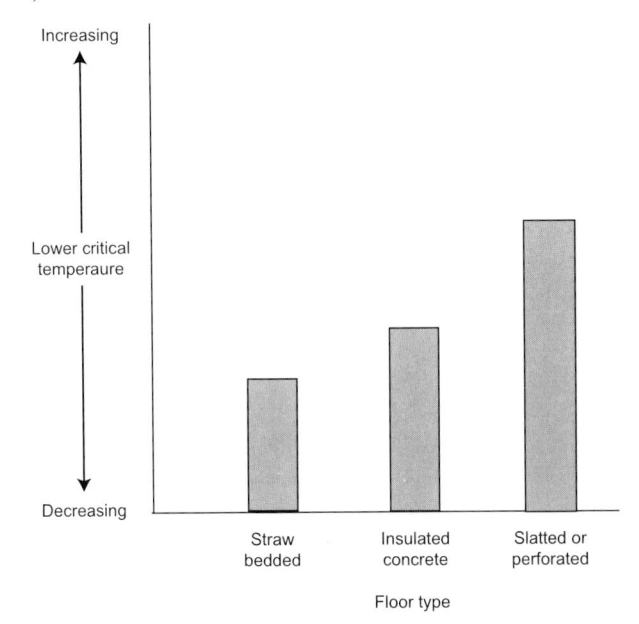

Figure 3.5 The impact of different floor types on lower critical temperature.

Straw acts as a thermal barrier and slows down the dissipation of heat (Colin Wooldridge).

A pig which is in direct contact with a solid or perforated floor inevitably loses more body heat through it. Straw, however, during conditions of very hot weather can actually reduce the Upper Critical Temperature. This arises because the bedding acts as a thermal barrier and so slows down the dissipation of heat through the actual floor surface. Wet bedding or wet flooring will increase the LCT because of the energy losses associated with evaporating moisture.

Although pigs are less sensitive to cold when housed on a generous straw bed, labour input can however be over-stretched and hygiene and health compromised on bedded systems. Slotted and unbedded solid floors offer less thermal comfort but labour demands are reduced and there are greater opportunities for improving hygiene.

The physics of heat loss through floors is governed by the thermal resistance of the floor and the area of body contact with it. Poorly insulated floors enable much heat to be lost from the area of the pig's body in direct contact with the floor. If a floor is poorly insulated the room or kennel temperature will have to be higher to compensate for the heat lost through it, otherwise feed intake will increase. Pioneering work by Bruce (1979) enabled the prediction of likely heat losses through various floor types.

Table 3.1 Impact of floor type on lower critical temperature

Weight of pig (kg)	Feed intake	Dry straw bed °C	Insulated concrete °C	Punched metal °C
5	Low	27	28	29
20	High	15	16	20
60	High	9	11	16
140 (single sow)	Low	18	20	22

Weight of pig (lb)	Feed intake	Dry straw bed °F	Insulated concrete °F	Punched metal °F
11	Low	81	82.4	84
44	High	59	61	68
132	High	48	52	61
309 (single sow)	Low	64.4	68	71.6

Table 3.1 has been abstracted from Bruce's data and demonstrates that:

- Floor type has a big impact on LCT
- Increased feed levels can help mask the effects of a poorly insulated floor
- As pigs grow, the negative impact from poorly insulated floors is reduced.

Choosing and correctly managing a hygienic, durable and comfortable floor surface is a key factor in commercial pig production.

Air speed at pig level

Given the absence of fur or feather, it is not surprising that pigs should feel sensitive to draughts. If group housed pigs are kept just above their Lower Critical Temperature, once air speed reaches 0.15 metres/second (29ft 5in/minute), they will start to feel cold. Significantly, people can detect the cooling effect of such an air speed on the back of the hand, provided air temperature is below that of the skin surface.

Figure 3.6 provides an indication of the impact of increasing air speed on group-housed 30 kg (66 lb) weaners.

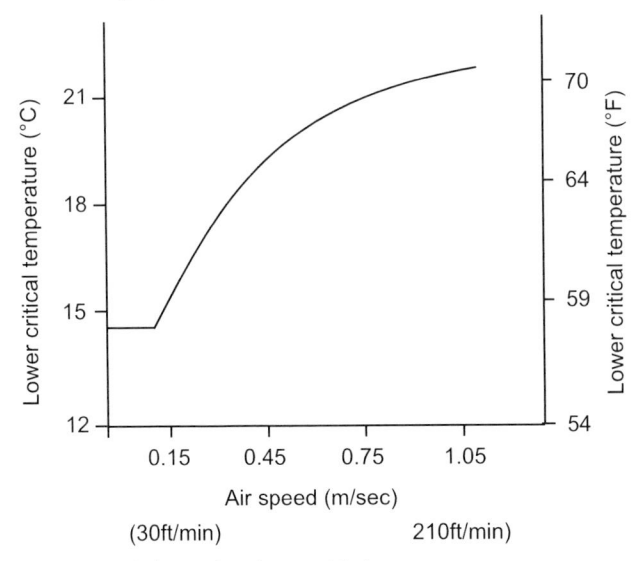

Figure 3.6 The impact of air speed on lower critical temperature.

The ability of draughts to create heat loss depends on body size, wetness of skin, type of floor and air temperature at pig level. Small pigs kept at low temperatures and subjected to draughts when lying on a wet, non-bedded floor are particularly vulnerable. During heat waves and conditions of maximum stocking density, high air speeds can be used to cool pigs. If air inlets, particularly during cold, windy weather, are not adequately baffled, cold air tends to surge downwards towards pigs and this can result in debilitating digestive and respiratory problems as well as an unhelpful dunging behaviour.

Relative humidity

There is a close relationship between humidity and temperature. Provided pigs are housed within their normal temperature range, i.e. within the comfort zone, there is a high degree of tolerance to a range of humidity levels. Pigs become challenged when humidity is high and conditions are hot. Evaporative cooling is impaired and the pig "re-writes the equation" by restricting its feed intake, i.e. it minimises production of more heat energy. Dirty pigs or a wallowing response often arises during hot, humid conditions. During cold conditions, high humidity levels are particularly unhelpful. Water droplets on the skin have to be evaporated from it, this requires dissipation of heat energy which makes already cold pigs feel even colder. Low humidity discourages the settlement of dust and dries out the pig's skin and its usually moist respiratory tract. A combination of dust and dry mucous membranes in the respiratory tract is a pre-disposing factor towards respiratory problems. If humidity is too high, this could be indicative of inadequate insulation. In poorly insulated buildings, surfaces tend to be cold and condensation can occur even at low humidity levels. A relative humidity level of 60 – 90 per cent is acceptable under most practical circumstances. In situations where humidity is beyond this range, pig attendants should assume that there are potentially serious shortcomings in the environment. Experienced pig people value the "glasses and nose" test. If, after a few minutes in a pig building spectacle wearers find their lenses continue to mist-up, the humidity level is probably excessive. Likewise, a dryness and slight burning sensation in the nose generally indicates that conditions are too dry.

Radiant effects

Heat loss from a pig is influenced by convection but this is exacerbated

Thermal Comfort, Light and Electrical Safety 79

when there is a direct radiant heat loss from the animal. Where there is a large temperature differential between an animal and its kennel or building, the radiant loss can be high. This is likely to be a key factor in a poorly insulated building; the worst insulated walls tend to 'steal' heat from the pigs. Therefore, under those conditions the LCT of the pigs would be increased.

Farrowing house creep heaters and those for first-stage weaner housing emit radiant heat. This makes pigs feel warmer than temperature readings imply. Whilst temperature readings are useful, the posture of the pig should also be observed since this gives the stockperson a more meaningful indication of pig comfort. Pigs huddle when they feel cold and maximise their surface area by spreading out when they feel warm.

Radiant heat makes pigs feel warmer than temperature readings imply (Quality Equipment/ BHR Communications).

Group size

Just as large pigs have a lower ratio of surface area to liveweight, and so lose proportionately less heat, the same principle applies to group size.

80 *Pig Environment Problems*

When pigs lie together proportionately less area of skin is exposed, hence less heat is lost. This is the reason why group housed pigs have a lower critical temperature than those housed singly. An individually housed sow on low feed intake would start to feel cold at 20°C, whereas if the sow were in a group of five, the LCT would be down to around 16°C. On commercial farms a balance must be struck between the desirability of furnishing pigs with more overall living space and the need to provide them with a thermally comfortable lying area.

On commercial farms a balance must be struck between the desirability of furnishing pigs with more overall living space and the need to provide them with a thermally comfortable lying area (JSR Genetics/BHR Communications).

The lighting regime

The Welfare of Farmed Animals (England) Regulations 2000 (S.I. 2000 No. 1870), Schedule 1, paragraphs 3 and 16 state that:

3. *"Where animals are kept in a building, adequate lighting (whether fixed or portable) shall be available to enable them to be thoroughly inspected at any time"*
16. *"Animals kept in buildings shall not be kept without an appropriate period of rest from artificial lighting"*.

The Welfare of Farmed Animals (England) (Amendment) Regulations 2003 (S.I. 2003 No. 299), Shedule 6, Part II, paragraph 8 states that:

8. *"Where pigs are kept in an artificially lit building then lighting with an intensity of at least 40 lux shall be provided for a minimum period of 8 hours per day subject to paragraph 16 of Schedule 1 to these regulations"*

(DEFRA, 2003)

Given that pigs evolved in woodland conditions, i.e. under a tree canopy, it is perhaps not surprising that lighting requirements have relatively little impact on pig comfort. However, those who care for pigs must have facilities for observing and closely inspecting their pigs. Section 59 of the UK Pig Welfare Code suggests: *"You should have enough fixed or portable light fittings available at any time if you need to inspect any animals, e.g. during farrowing."* Routine tasks are also more efficiently undertaken under effective lighting conditions. Work routines are easier and quicker and often staff morale is boosted in well lit environments.

The 'lux' is the unit used to measure light intensity. A lux level of 80,000 is typical of a bright summer's day whereas lighting in a modern office provides around 500 lux. Significantly, the latter coincides with the degree of light intensity required to examine pigs inside a building. The 500 lux recommendation is based on a standard which takes account of the average illuminances over the given area throughout the lifetime of the installation. Light fittings coated with dust emit reduced light levels and so the degree of illuminance actually provided must make allowances for this reduction. Whereas a light-cleaning regime is invariably part of the hygiene protocol on all-in/all-out systems of production, sometimes dust build-up is allowed to impair lighting efficiency on continuous throughput systems. In windowless buildings a light intensity providing at least 50 lux is required at all times. Windows coated with dust impair light penetration.

There is no evidence that increased lighting leads to an improvement in growth rates and feed conversion efficiency in growing pigs. Limited evidence supports the principle of providing at least 400 lux for 16 hours a day for breeding stock. In some instances it has been shown to reduce days to oestrus, particularly in gilts and there are also indications that extra light during lactation stimulates milk production.

Some pig farmers have reservations that keeping pigs in a well lit environment leads to hyperactivity and vices. However, it is more likely that this is simply an indication that the environment is too sterile in any event and needs enriching. Particularly in Scandinavia there is a trend to providing high levels of natural light in pig buildings. Anecdotal evidence suggests that the pigs adjust to the enhanced lighting regime and that farm staff respond positively.

When a building is first wired, it is helpful if any areas needing occasional extra lighting are identified and switched independently. Particularly with the trend to all-in/all-out production and imposition of a "white-glove clean" hygiene regime between batches, ensuring that lighting is dust and moisture resistant is likely to become more important. This requirement also has health and safety implications.

Electrical safety in a challenging environment

Lighting, as well as other electrical fittings in pig buildings, must be of sufficient durability to allow efficient functioning within the hostile pig environment. Within the actual pig building there is every likelihood that noxious gases, dust, moisture and impact damage will conspire to compromise the workings of any electrical devices meant to improve the environment. The subject of electrical safety within this harsh environment must also be considered. It is therefore prudent to ensure that only specialist equipment designed and manufactured to operate in this unique, challenging environment is used. Siting sophisticated control panels within the same air space as the pigs is not a good idea. Lighting can also be particularly vulnerable.

Pig farms in the United Kingdom must not violate The Electrical Safety at Works Regulations, 1989. In so far as they relate to matters under their control, the Regulations impose responsibilities on employers, employees and the self-employed.

The likelihood of impact damage or moisture and dust ingress is reduced if controllers and equipment have the appropriate Index of Protection (I.P.) rating. The degree of protection which electrical fittings have against the ingress of water and solid objects is known as the I.P. (Index of Protection) rating.

Control panel sited in a separate air space to that of the pigs (TXU Europe, formerly Eastern Group).

Table 3.2 Index of protection of electrical fittings

I.P. number	Solid objects	Water ingress
0	No protection	No protection
1	Accidental touch by hands (50mm)	Vertically falling water
2	Finger proof (12mm)	Water sprays up to 15° from vertical
3	Tool and wire proof (2.5mm)	Water sprays up to 60° from vertical
4	Small wire proof (1.0mm)	Water splashes from all directions – limited ingress
5	Dust protection – limited ingress	Low pressure water jets – limited ingress
6	Complete dust protection	Strong water jets – limited ingress

N.B.   25mm = 1 inch

Table 3.2 indicates that, unless the I.P. rating is at least 5 for solid objects, dust invasion would be a problem. Given the more widespread use of impervious cladding materials in pig buildings and the need for all-in/all-out production, avoidance of moisture penetration into the electrics is becoming more important. An IP of 4 only ensures resistance from water splashes whereas levels of 5 and 6 are required to cope with jets of water. These days I.P. 5.4 should be regarded as the very minimum degree of protection for light fittings and staff must be made aware of its limitations. Expert advice should be sought regarding the details of electrical safety.

4

THE IMPACT OF NOXIOUS GASES

Although carbon dioxide is the major waste product of the pig's respiratory system, within pig buildings noxious gases mainly arise because of manure breakdown. Changes in the microbial composition of pig faeces and urine give rise to chemical changes involving the discharge of noxious gases. Ammonia, carbon dioxide, carbon monoxide and hydrogen sulphide are the most common waste gases mainly associated with pig keeping. Whilst these gases are relatively stable, there are also other gases which are transitory breakdown products of anaerobic activity. The release of noxious gases arises because a vast range of anaerobic bacteria use protein, carbohydrates and fats present in manure as a source of energy. Organic acids, alcohols, aldehydes, amides, amines and sulphides are likely to be present under conditions of anaerobic decay. The degree of breakdown is sensitive to moisture content, pH and temperature.

Gases with a low molecular weight exert a higher vapour pressure than more complex heavier gases. This results in low molecular weight gases being more volatile and this increases the likelihood of polluting the atmosphere. Solubility of gases also influences that potential for aerial pollution since they tend to become trapped in solution and react with other retained substances.

Provided environments are adequately ventilated, gases such as ammonia, carbon dioxide, carbon monoxide and hydrogen sulphide are generally not a problem to pigs or people. There is, however, a risk of lethal poisoning if anaerobic pig slurry is pumped or stirred in an under-ventilated air space (Taiganides and White, 1969).

Ammonia

Pigs use dietary protein to maintain their bodies and to fuel various productive processes, e.g. growth and lactation. Nutritionists endeavour

86 Pig Environment Problems

to provide the right type of protein by specifying a blend of essential amino acids. Nitrogen is a key component of dietary amino acids, some is retained within the body but there is only a limited facility for storage. Surplus nitrogen must therefore be excreted from the pig's body. Much of this loss is in the form of ammonia gas which is released from both liquid and solid manures. Other nitrogen rich gases such as amines, nitrous oxides and nitric oxide also are released from manures. Nitrogen can also be released as ammonium compounds in an aerosol form into the atmosphere from feed, skin or manure fragments (Figure 4.1).

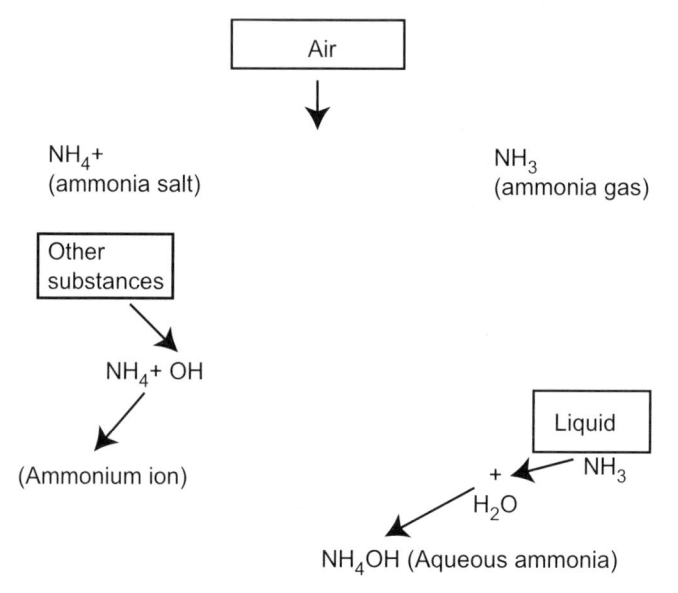

```
                      ┌─────┐
                      │ Air │
                      └──┬──┘
                         ↓
  $NH_4^+$                           $NH_3$
  (ammonia salt)                     (ammonia gas)

  ┌──────────┐
  │  Other   │
  │substances│
  └────┬─────┘
       ↘
      $NH_4^+ + OH$
       ↙                          ┌────────┐
                                  │ Liquid │
                                  └───┬────┘
  (Ammonium ion)              +  ←  $NH_3$
                             $H_2O$
                              ↙
              $NH_4OH$ (Aqueous ammonia)
```

Figure 4.1 Nitrogen in ammonium compounds.

Since it is acknowledged that over 80 per cent of ammonia emissions arise from agriculture, it follows, therefore, that farmers will be under much pressure to reduce the level of nitrogen rich protein in animal feeds. Not only is ammonia (NH_3) present as a gas in the air, it is very soluble in water and forms aqueous ammonia where it is present as the ammonium (NH_4^+) ion.

The ammonium ion is also able to react with other substances resulting in the further discharge of ammonia gas.

In the process of de-amination which takes place in the liver and kidneys, oxygen reacts with amino acids to form uric acid. This substance does not contain ammonium as such and forms urea which is ultimately excreted through the kidneys. However, urea can be transformed into ammonium in the presence of the enzyme urease (Figure 4.2)

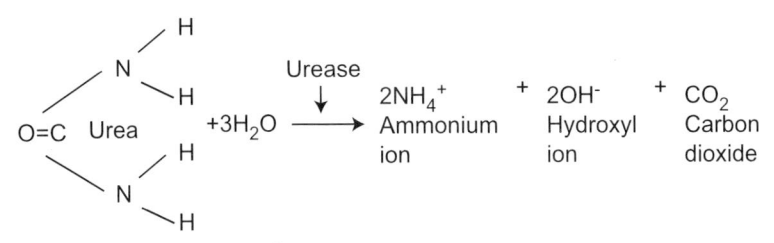

$$\text{Urea} + 3H_2O \xrightarrow{\text{Urease}} 2NH_4^+ + 2OH^- + CO_2$$
$$\text{Ammonium ion} \quad \text{Hydroxyl ion} \quad \text{Carbon dioxide}$$

Figure 4.2 Enzymic breakdown of urea.

Urease

The enzyme urease is present in faeces whilst most of the urea is in the urine. Hence mixing of the two facilitates the production of ammonium ions as outlined in the foregoing equation. The rate of production is enhanced when urea concentration is strongest, particularly under alkaline conditions and at elevated temperatures. The latter helps explain the reason why odours from pig units are most problematic during heatwaves.

Under anaerobic conditions nitrogen-rich compounds in manure can be transformed by microbes into carbon dioxide, methane, hydrogen, hydrogen sulphide, ammonia, amines, indoles, skatole and butyric acid (Oldeburg, 1989). The associated pathways are shown schematically in Figure 4.3.

In the presence of oxygen, nitrogenous substances in manure are also broken down in the heat-generating process of composting. Ammonia gas is the fate of most of this nitrogen, some of it escapes in the air, whilst some is incorporated into microbial protein. Alternatively, in the presence of oxygen, ammonia can be nitrified, ie. converted into nitrates. Once all the oxygen has been used up, the anaerobic process of denitrification is triggered, whereby nitrate is converted into nitrogen gas.

Given this basic understanding of the characteristics of pig waste, there are several practical measures that can be undertaken to lessen the pollution risk. Early separation of urine and faeces will reduce the rate of ammonia volatilisation. The need for "clean" and "dirty" areas on a pig unit is important since spillages of pig waste tend to increase its surface area and in the presence of oxygen create more ammonia. Storing manure at temperatures below 10°C markedly reduces ammonia emissions as does the restriction of air movement directly above the manure store. Hence there is a move towards roofed muck heaps and lidded slurry stores.

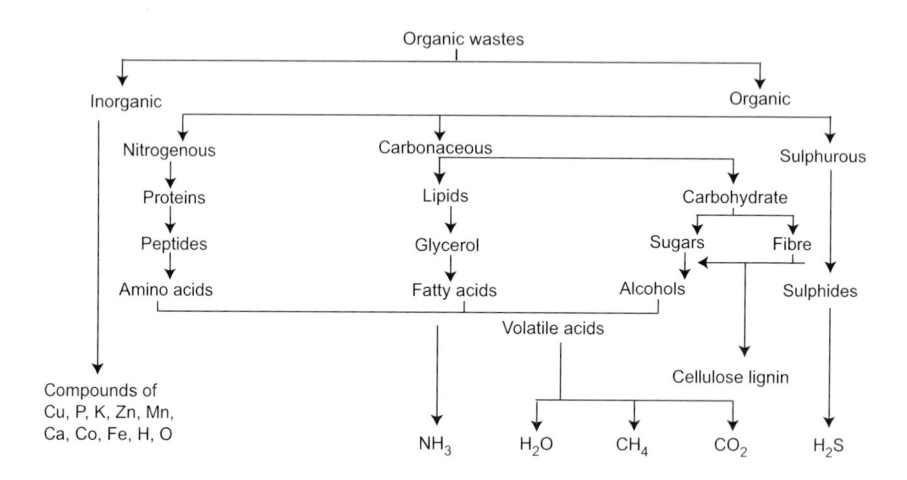

```
                          Organic wastes
            ┌──────────────────┴──────────────────┐
         Inorganic                              Organic
            │           ┌────────────────────────┼──────────────┐
        Nitrogenous  Carbonaceous                            Sulphurous
            │            │                                       │
         Proteins      Lipids                  Carbohydrate
            │            │                      │
         Peptides     Glycerol              Sugars ←──── Fibre
            │            │                      │
        Amino acids   Fatty acids           Alcohols          Sulphides
            │            └────────┬─────────────┘
            │                Volatile acids
            │                     │          Cellulose lignin
     Compounds of                 │                │
     Cu, P, K, Zn, Mn,            │                │
     Ca, Co, Fe, H, O            NH₃   H₂O   CH₄   CO₂         H₂S
```

Figure 4.3 A schematic representation of anaerobic decomposition processes in manure (Taiganides, 1987).

The incorporation of a below-slat slurry–scraper even in the absence of solid separation will help reduce the production of ammonia gas. Dutch pig farmers appreciate the benefits from using below-slat pans and gutters into which slurry is directed. This effectively reduces the surface area from which ammonia can be emitted. Provision of a shallow storage pit is also helpful in that it reduces the amount of time in which slurry is stored below slat level. Alternatively, pipes can be used to convey slurry away from pig buildings for storage elsewhere. Automatic flushing mechanisms over almost level slurry channels also helps reduce ammonia. Within pig buildings, design features such as slopes and steps can be used to create clean lying areas and cooler, wetter or draughtier dunging areas. On solid muck systems daily or twice daily mucking-out markedly reduces ammonia levels. Provision of a deep straw bed within a well ventilated vast air space also encourages pig waste to compost and the heat produced evaporates the trapped water.

Figure 4.4 demonstrates that ammonia concentration rapidly diminishes with height since production and storage of pig waste takes place at ground level or lower.

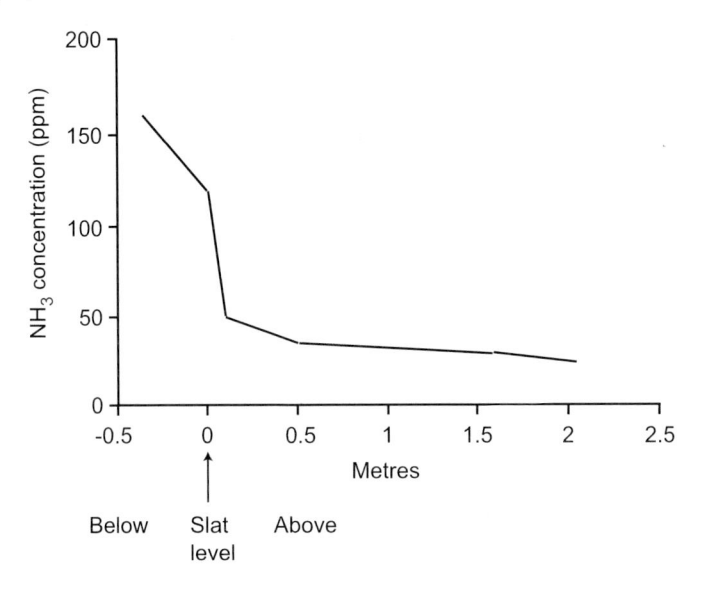

Figure 4.4 Concentration of ammonia at different heights above the manure in a pig building in which manure was 0.4m below slat level. (source: International Commission of Agricultural Engineering)

Ammonia acts as an irritant to the respiratory tract of both pigs and people. Trials have indicated no significant deterioration in the feed conversion efficiency of pigs when subjected to levels of 10, 50, 100 and 150 ppm although feed intake and daily gain were significantly reduced (Stombaugh *et al.*, 1969). Trials in America, however, have indicated a negative impact with increased ammonia levels in weaners already suffering respiratory stress (Drummond *et al.*, 1981). The practical significance is the negative impact that high levels of ammonia in combination with dust and microorganisms can have on respiratory health. If other aspects of the environment are sub-optimal, on commercial farms ammonia could be harmful to pigs at lower levels than those in which ammonia has been tested under controlled experimental conditions.

Carbon dioxide

Carbon dioxide is a by-product of the metabolism of the pig and can be used to indicate air quality within a building. Since it is a heavy gas, if measurements are taken at pig level, an accurate assessment of the quality of air actually experienced by the pig is possible. Carbon dioxide production arises from three different sources:

- As a by-product of respiration
- As a breakdown product of urea in urine
- As a waste product when microbes break down pig manure

Most carbon dioxide produced within pig buildings results from respiration. Since pigs must breathe, it follows that special efforts must be made to clear carbon dioxide from the vicinity of the pig and its macro-environment. This is effected by increasing the ventilation rate which must arise in circumstances that allows the pig to remain within its comfort zone, except in the event of an emergency.

The density of air is 1.29 kg/m³ (0.0805321 lb/ft³) whereas carbon dioxide has a density of 1.98 kg/m³ (0.1236074 lb/ft³). This difference governs the physics of air mixing. Within a pig building gases are being mixed all the time and the amount of heat production from the pig is a key factor. Pigs dissipate heat, this initially warms up the air immediately around them and since this is relatively light, it becomes upwardly mobile and is replaced by cooler air. This tends to fall into areas where there are fewer or no pigs, i.e. where there is less upward air movement, It can often account for "dead spots" within a pig building.

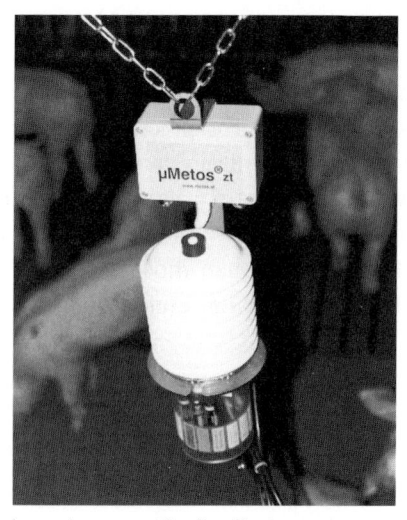

Data-logger monitoring pig environment (Quality Equipment).

Since carbon dioxide has a marked impact on air movement, it can be used as an indicator of air quality. Levels of the gas can be measured either continuously or intermittently but this is rarely undertaken in pig buildings because of the need for specialist equipment. During periods of intense activity such as during feeding time or when people enter a pig pen, carbon dioxide levels increase. Elevated carbon dioxide levels also tend to be associated with environments in which pigs have become agitated during or following a period of tail-biting.

Carbon dioxide concentration within the atmosphere is 0.034% (340 ppm) and this markedly increases to 5% (50,000 ppm) in exhaled breath. Within a pig building a typical target level for carbon dioxide at the minimum ventilation rate is 0.3% (3,000 ppm) and 0.5% (5,000 ppm) is a practical upper limit. This is why carbon dioxide can be used as an indicator of air quality. Higher levels of carbon dioxide can be tolerated without impairment of pig health but prove less than comfortable to people. Elevated levels of carbon dioxide tend to suppress pig appetite and make air spaces feel "heavy". In sophisticated ventilation systems with a facility to re-circulate air, there is an increased risk of imposing higher than ideal levels of carbon dioxide unless careful attention is paid to minimum ventilation rate settings.

Hydrogen sulphide

This is the most lethal gas ever likely to be encountered by pigs and people within a pig building. Hydrogen sulphide is usually produced when slurry is stored under anaerobic conditions, hence pit depth and storage times are key factors. This lethal substance is actually stored as gas bubbles trapped within the liquid manure. When the manure is agitated, e.g. by a slurry tanker pump, the gas is released. Hence it is vital that on all pig units there must be a well-understood safety policy regarding slurry management. This must safeguard the respiratory health of the pigs and ensure that personnel leave pig buildings prior to the commencement of slurry agitation. A submersible re-circulation pipe should be used when mixing slurry under slats and backflow of gas must be prevented.

Figure 4.5 indicates that unless steps are taken to step-up the ventilation rate in a pig building, there will be a rapid build-up of hydrogen sulphide during slurry emptying procedures.

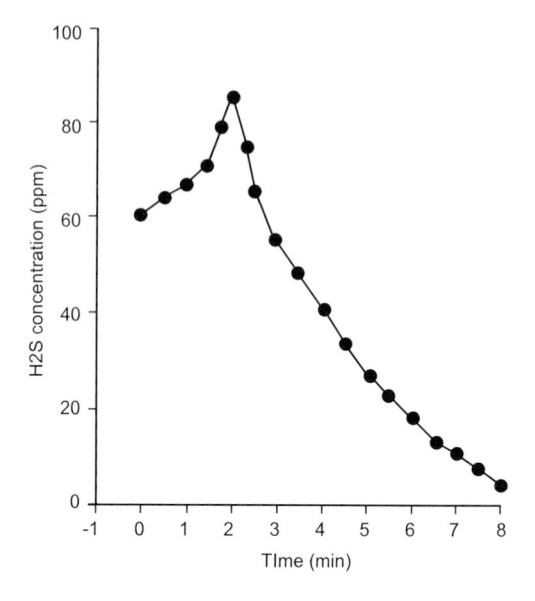

Figure 4.5 Hydrogen sulphide concentration at pig level adjacent to the sluice gate when emptying the slurry channel in an unventilated building (Robertson, 1971).

As slurry mixing continues, the rate of gas removal slows down. Hydrogen sulphide tends to circulate along the same routes as natural air movement patterns.

Table 4.1 indicates that a concentration of 0.01 to 0.7 ppm, can be detected, but hydrogen sulphide only becomes characterised by the foul smell of rotten eggs when levels exceed 3 ppm. If concentrations are allowed to exceed 500 ppm, the consequences are very serious. People, however, seem to be troubled by hydrogen sulphide more than pigs. Levels of 10 ppm result in eye irritation in most people and this must be regarded as the upper limit. Levels of 20 ppm would suppress appetite in pigs and be stressful to people.

Carbon monoxide

If space heaters or piglet creep heaters powered by gas are used in pig buildings, there is a risk of carbon monoxide build-up. It arises during conditions of incomplete combustion of gas heaters and is most likely to

Table 4.1 Physiological response of adult humans to hydrogen sulphide (Nordström and McQuilty, 1976).

Effect	Gas concentration (ppm)
Least detectable odour	0.01-0.7
Offensive odour (bad eggs)	3-5
Eye irritation	10
Irritation of mucous membranes and lungs	20
Irritation of eyes and respiratory tract (1 hour)	50-100
Olfactory – nerve paralysis, fatal in 8-48 hours	150
Headaches, dizziness (1 hour), impairment of nervous system	200
Nausea, excitement, insomnia, unconsciousness, possible death (30 minutes)	500-600
Respiratory paralysis, death	700-2000

occur when heater maintenance is inadequate and ventilation rate is severely restricted (Wood, 1979). Dusty, under-ventilated kennels or compact, draughtless creep boxes would be the most likely susceptible environments. The gas is toxic to both humans and pigs since it rapidly forms a stable compound, carboxyhaemoglobin, when drawn into contact with haemoglobin. McAllister (1966) undertook trials in which carbon monoxide levels of 4,000 ppm resulted in the deaths of adult pigs. There have been incidences of carbon monoxide causing stillbirths on commercial farms at gas concentrations between 150-200 ppm.

5

DUST – A PROBLEM FOR PIGS AND PEOPLE

The components of dust

Dust is made up from dry solid particles of feed, skin, hair, bedding and faeces and is generated mechanically. Since pigs housed indoors are usually kept in a dry environment, this results in the generation of dust from various physical processes, this is shown schematically in Figure 5.1.

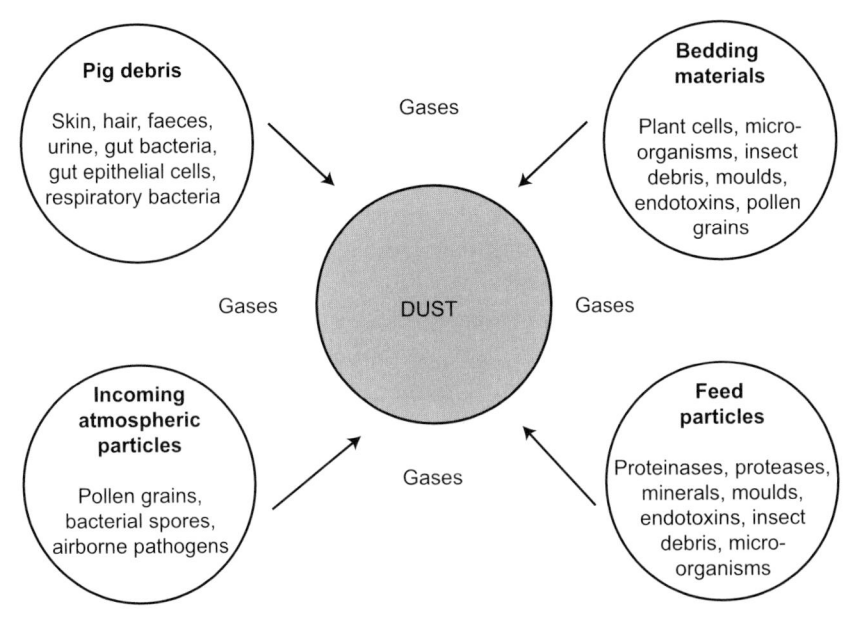

Figure 5.1 The diversity of dust.

Whereas some dust settles, airborne dust along with gases, microorganisms, water vapour and other substances forms part of the aerosol component within a pig building and acts as a carrier for all these substances. Microorganisms make up less than 1% of these airborne particles but they have a big influence on disease transference and are also often responsible for allergies.

Kennels can be a source of dust. (Farmex)

Particle size

The largest dust particles are known as inspirable particles. Whilst these are a source of irritation in healthy pigs and humans, the nose acts as an efficient filter and so limits the impact. Finer, respirable particles get through the trachea and bronchi to the alveoli in the lungs and they can cause temporary and permanent lung damage.

There are no actual regulations about the maximum permissible levels of dust breathed in by pigs, but there are upper levels for humans. In the United Kingdom the Maximum Exposure Limit (MEL) of dust to humans is time based. The upper limits are:

Inspirable dust $10mg/m^3/8$ hours - Deposited in the respiratory tract via the nose and mouth

Respirable dust $4mg/m^3/8$ hours - Deposited in the gaseous exchange region of the lungs

These thresholds are set under the Control of Substances Hazardous to Health (COSHH) Regulations 1988 (Smith, 1991).

Dust particles exhibit various shapes and sizes which, along with density, govern their settling velocity. Within ambient air, respirable dust is defined as 100% of all 0.1µm particles, decreasing to approximately 40% of 5µm particles and containing no larger 10µm diameter particles.

Figure 5.2 provides details of the size ranges of the many different types of dust.

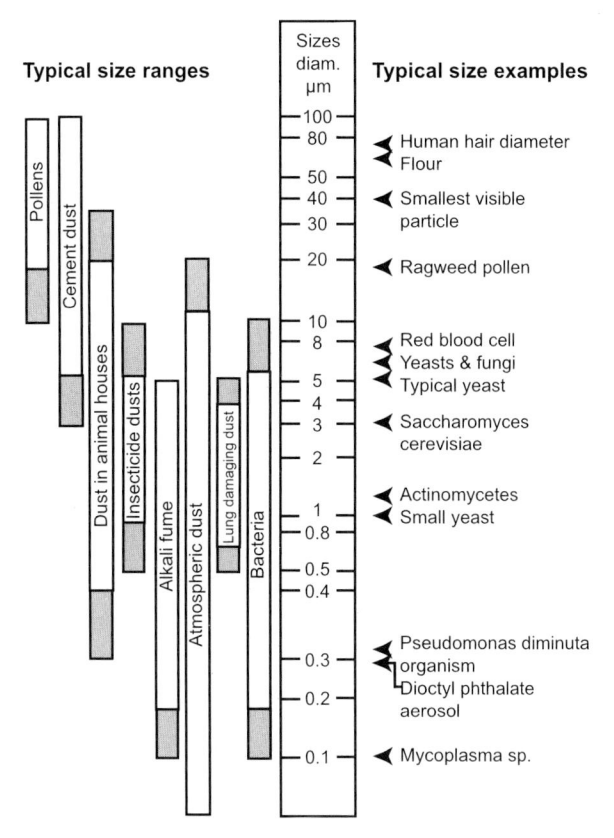

Figure 5.2 Size ranges of different types of dust.

In particular, it indicates the range of dust particle size in animal houses, those that cause respiratory disease and highlights the fineness of Mycoplasma bacteria,

Practical problems with dust

Figure 5.3 outlines typical concentrations of airborne dust within a pig building.

98 Pig Environment Problems

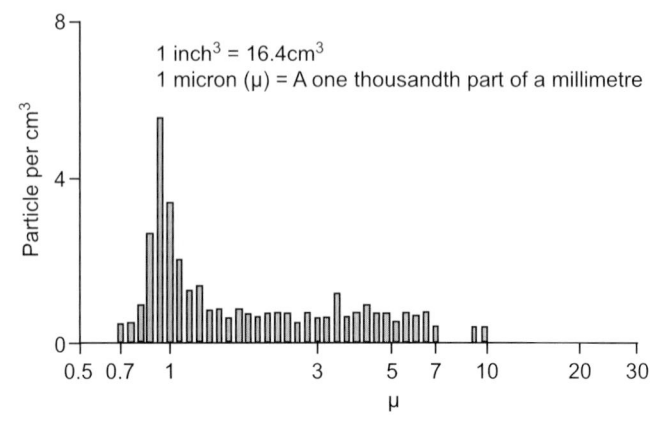

Figure 5.3 The concentration of different particle sizes in pig house air.

Some dust enters a building in the incoming air and its concentration builds up from feed, bedding and the pigs themselves. Although only negligible amounts of dust invade pig buildings via incoming air, this can be a major source of pathogens and allergens. Fine, disease bearing dust particles are carried in the wind. Hence, in order to avoid the transmission of respiratory diseases between farms it is important to heed veterinary guidelines about siting new pig units too close to existing units and think carefully about layout options when new buildings are added to an existing pig farm.

Dust particles with the highest diameter and density settle more quickly inside a building. As a result of gravity they fall to the floor, land on bedding or coat fixtures and fittings as well as any pigs and people within the building.

Table 5.1 provides details of the settling velocity of dust of different diameters.

In the absence of fan ventilation, heat output from the pigs will cause air movement at around 0.05 metres/second or more. Table 5.1 indicates that dust particles with only 1μm diameter in free air would settle at 0.03 metres/second, whereas heavier particles would fall much more quickly. Hence, although fans might not be used, thermal output alone will keep fine dust airborne above the pigs.

Table 5.1 Settling velocity of dust particles of different sizes

Diameter (μm)	Settling velocity (m/sec)
1	0.03
2	0.12
3	0.26
4	0.47
5	0.8
10	3.3
20	12

(International Commission of Agricultural Engineering)

Fan ventilation provides an opportunity for dust removal in exhaust air which has the same dust concentration per unit volume as the air within the building. Since increasing the ventilation rate increases the amount of air coming into a building, it would first seem that this must proportionately decrease dust concentration because of the dilution effect. However, the situation is more complicated in practice. If ventilation rate is increased, air speed within the building increases and this tends to disturb settled dust. Furthermore, increasing the ventilation rate reduces room temperature. This means that ventilation is a rather limited mechanism for the control of dust.

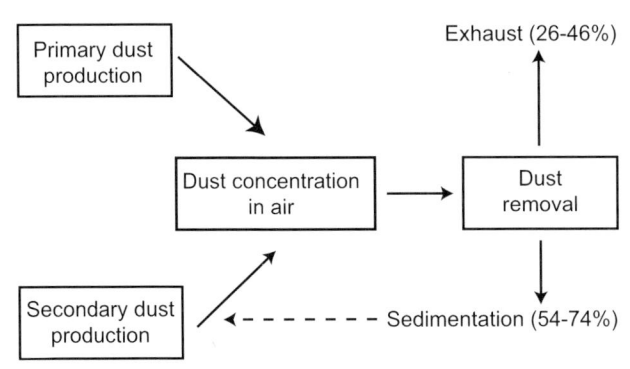

Figure 5.4 Dust production and removal.

Figure 5.4 represents a simple model describing dust production and its removal. Whilst dust accumulation on surfaces looks untidy and could

100   *Pig Environment Problems*

be harbouring disease, pigs and people find airborne dust more troublesome. Any strategy for removing this dust will involve minimising the impact of dust removal itself.

The likelihood of reducing the level of airborne dust increases when different removal strategies are used in combination. First, however, the individual sources of dust must be identified. Thereafter, the contents can be assessed to determine the potential biological activity and impact on pig and human health. Once the source of dust has been defined, it is then possible to identify the most suitable apparatus for undertaking quantitative measurement.

Given this information, tackling the dust problem will depend on how much is present and whether or not it is a serious health hazard to pigs and people. Identical systems sometimes result in different dust burdens because of the management strategy adopted. If a pig building is particularly dusty, questions must be asked whether or not any management shortcomings are responsible. Only when all these factors have been considered will it be possible to work out the most appropriate dust removal strategies, bearing in mind the cost of implementing them. Figure 5.5 summarises the key stages in a dust control policy.

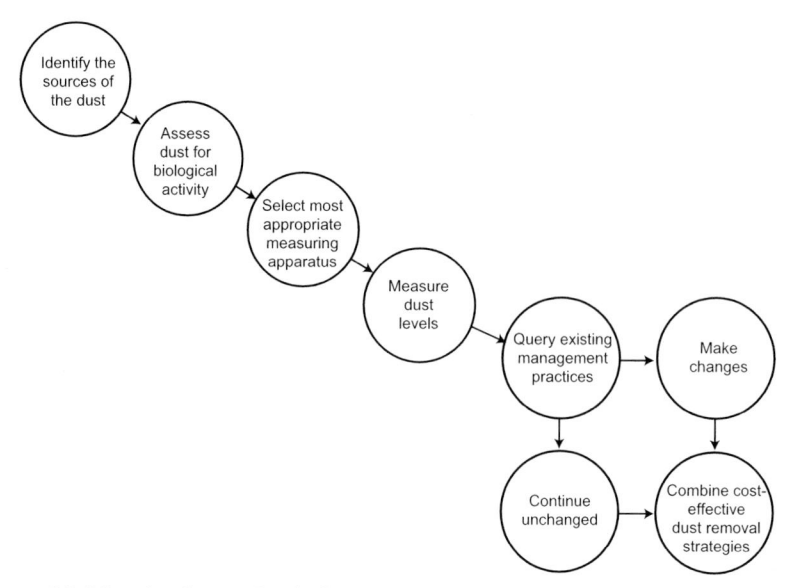

Figure 5.5 Managing dust on the pig farm.

It must be understood that there are many existing commercial situations where there is little scope for *reducing* dust levels. In such circumstances the focus must be on *removing* the dust as effectively and quickly as possible.

As Figure 5.4 indicates, secondary dust can be a problem. This is the dust which has previously settled but when stirred-up becomes airborne again. This secondary dust can be created by pig activity, vibration of mechanical equipment or as a result of specific airflow patterns Air speed is also a crucial factor; when air velocity is high, settled dust can be disturbed and re-entrained into the air system (Robertson, 1989). This difficulty is associated with any ventilation system that actively controls airflow by using high inlet velocities, such as polythene duct, high speed jet or controlled inlet.

If grain is harvested or stored under very dry conditions, dust levels tend to be higher. Damp harvesting or inadequate storage can lead to mite infestation or the production of moulds and toxins which can seriously impair respiratory health. Larsson (1983) demonstrated that if moisture content were higher prior to grinding or crushing, dust levels could be reduced by up to 78 and 82 % respectively. Fat spraying pig feed not only boosts the energy content but it also reduces dust incidence. Heber and Stroik (1988) demonstrated in laboratory trials that dust levels could be reduced by 76-99% when 0.5,1.0 and 2.0% oil was added to pig feed. Pelleting pig feed or presenting it in a wet or moist form can help reduce dust production. Excessive dust production is a major reason why floor feeding of meal has no place in modern pig production. A large survey involving finishing pigs (Baekbo, 1989) indicated total dust levels around 1.6 mg/M^3 on wet fed systems compared to 2.6 mg/m^3 with dry feeding. When pigs are kept on a restricted feeding regime, they tend to become agitated before, during and after feeding and this leads to an increase in physical interactions which creates dust. In a six month long study involving 20 commercial pig buildings, mean total dust concentrations reached 4.6 mg/m^3 on ad-lib feeding systems, whereas on restrict-fed systems dust levels averaged 10.0 mg/m^3 (Robertson, 1992).

A study undertaken by Wathes *et al.* (2003) at BPEX's Stotfold Pig Development Unit involved nine hundred and sixty weaner pigs exposed for 5½ weeks to controlled concentrations of airborne dust and ammonia

102 Pig Environment Problems

by continuously injecting incoming air with these pollutants. The results indicated that inhalable dust concentrations of 5.1 to 9.9 mg/M^3 combined with ammonia concentrations of up to 37 ppm adversely affected pig performance. The researchers emphasised the need for commercial pig producers to prevent the build-up of dust and ammonia in pig buildings so that respiratory health would not be impaired.

Dust and gas monitoring at Stotfold Pig Development Unit (BPEX).

Modern feed hoppers and feed distribution systems should be designed to limit dust production. The trend is towards the distribution of feed within sealed systems and discharge on demand only in small portions, often dampened with an integrated watering facility.

The cubic capacity of pig buildings can have an influence on dust levels. In low compact buildings there is a tendency to under-ventilate in order to avoid causing draughts at pig level. In particular, this can lead to an accumulation of dust on surfaces. Kennel environments can become very dusty even when located within vast well-ventilated air spaces. Practising continuous throughput, especially in finishing houses on breeder/feeder pig units eventually leads to an unacceptable build-up of dust since there is no opportunity to break the cycle. Elimination and reduction of settled dust is a major bonus associated with depopulation and cleaning of all-in/all-out production systems.

When pigs become over-crowded or if they have insufficient hopper space, competition between pigs increases. They become restless and their increased activity creates more dust. When pigs are disturbed because of intervention by the stockperson, dust levels tend to increase. Particularly on restrict-fed systems, pigs associate people with feed, human presence therefore agitates the pigs and this creates dust. Anticipation of feeding time can be a problem on both manually and automatically controlled systems when feed is restricted. Hence, dust levels tend to vary throughout the day on account of different degrees of pig activity. Nilsson (1982) recorded a five-fold elevation in dust levels just before feeding and 3.5 times higher levels during feeding compared to daily averages. Research and practical experience indicate a wide variation in dust production levels, not only between different buildings but between pens within the same building and from day to day.

The dustiness of bedding influences airborne dust concentration. Particularly in bedded farrowing pens, there is a trend towards the use of dust-extracted chopped straw or dust-extracted softwood shavings. Well designed slatted floors enable dust particles to fall through the slotted area into the slurry channel. However, in situations when slat width is particularly wide and humidity low, dry solid waste can accumulate on the slats and be ground into dust by the physical activity of the pigs.

Well designed slatted floors enable dust particles to fall through the slats. (BPEX)

104   *Pig Environment Problems*

Straw bedded pigs at Stotfold Pig Development Unit (BPEX).

The degree of relative humidity, ambient temperature and type of ventilation system can have a marked influence on dust removal. A high relative humidity tends to reduce dust levels, when temperatures are low, ventilation rates are lowered, hence less air is put through the building and this allows more dust to accumulate. In naturally ventilated buildings there is little control over air movement, hence dust tends to accumulate more than in fan ventilated systems.

Well designed sealed feeding systems help minimise dust generation (Simco Systems).

Figure 5.6 highlights the difficulties associated with measuring dust and outlines the questions which must be asked.

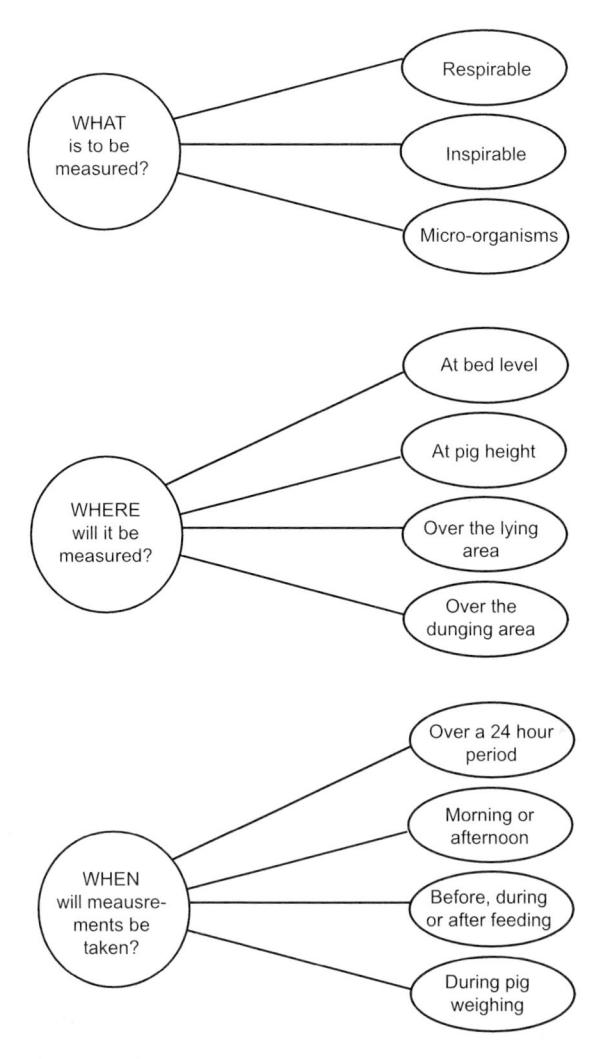

Figure 5.6 Measuring dust in pig buildings - the key questions.

WHAT is to be measured?
- Respirable
- Inspirable
- Micro-organisms

WHERE will it be measured?
- At bed level
- At pig height
- Over the lying area
- Over the dunging area

WHEN will meausrements be taken?
- Over a 24 hour period
- Morning or afternoon
- Before, during or after feeding
- During pig weighing

Readings are more meaningful when the location of sampling points are defined. The validity of readings increases when the number of samples taken reflects the degree of variability within the particular building tested. Single measurements taken in isolation are meaningless and must be related to specific work routines or averaged over a typical 24 hour period, e.g. a reading taken during a short period of "brushing operations" when the

106 Pig Environment Problems

dust is disturbed, would differ markedly from a reading taken during still air conditions. Readings taken when pigs are agitated because of the unexpected presence of a technician are likely to be higher than when pigs are left undisturbed.

Various methods and apparatus are available for measuring dust in pig buildings. The cassette method which involves pumping air through a filter paper is popular. This demands accurate calibration of the flow rate of air through a cyclone pump, recording the duration of the test plus "before and after" weighings of the 25mm (1 inch) diameter filter paper. Another dust measuring device is the optical particle counter. This works on photo-electric principles and provides continuing readings of dust at a given point. Whilst this method is not suitable for recording absolute values, it can be used to track relative levels of dust.

The best strategy for dust reduction is to design pig buildings and feeding systems so that the creation of dust is minimised. Dust reduction techniques usually involve either increasing the sedimentation of dust or decreasing its re-circulation. Reduction strategies based on air filtration demand regular replacement of filters. Fogging with water has produced variable results whilst fogging with rape seed oil has proved more promising.

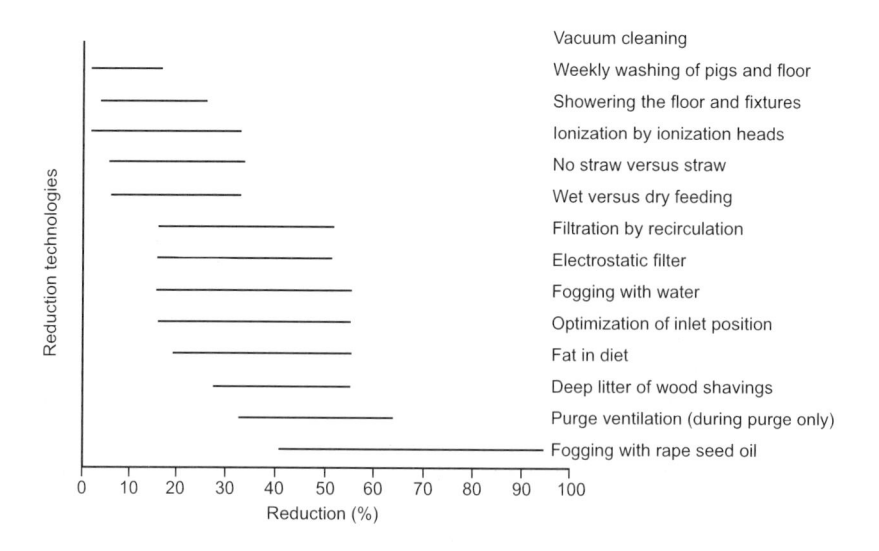

Figure 5.7 Expected reductions in airborne dust with different dust reduction technologies in livestock buildings. (source: International Commission of Agricultural Engineering)

Dust - A Problem for Pigs and People 107

Ionisation techniques have also met with some success in reducing dust levels.

Pearson (1989) undertook experiments on behalf of ADAS Farm Buildings Development Centre which involved the wet scrubbing of re-circulated air in first-stage weaner rooms (Figure 5.8).

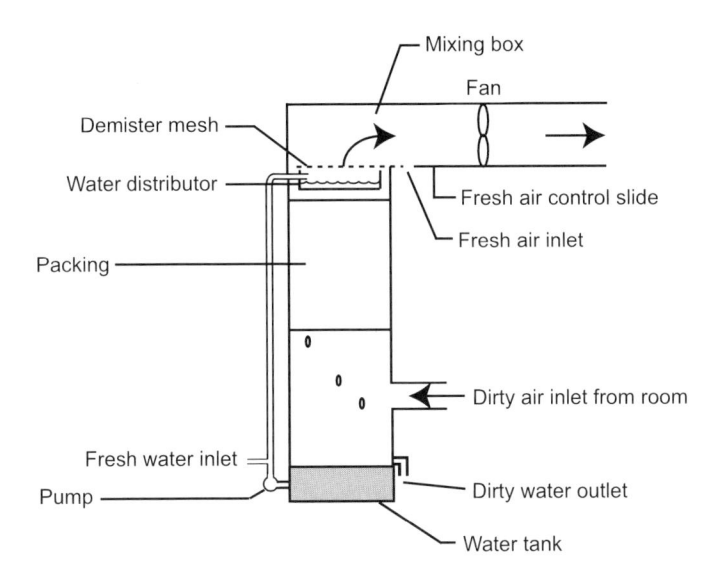

Figure 5.8 Design principles of a wet scrubber.

The concentration of total dust at the scrubber outlet was reduced to 0.7 mg/m^3. However, since the scrubber capacity only represented 11% of the maximum ventilation rate, dust levels within the actual room were only reduced from 4 mg/m^3 to 3.7 mg/m^3. Ammonia concentration was, however, reduced by about a third and there was also an overall reduction in odour emissions, but, overall, the process of wet scrubbing proved prohibitively expensive. The summary data presented in Figure 5.7 indicates that various dust reduction strategies have met with limited success and the problems associated with dust within pig buildings remain a major challenge for the pig industry. Within Europe there are indications of commercial interest in research projects which aim to measure coughing associated with increased dust and ammonia levels. Software is being developed that would enable pig keepers to be informed if threshold values were exceeded.

6

PRACTICAL CONSIDERATIONS

Calculations for comfort

Pig building designers, ventilation and heating engineers make use of some key mathematical formulae which facilitate fundamental calculations regarding the specification of the pig environment. These are shown in Table 6.1.

Table 6.1 The mathematics of pig environment.

Maximum ventilation rate (m^3/hr)	=	$\dfrac{\text{Sensible heat output from the pigs (W)}}{\text{Difference in temperature (°C) between inside and outside} \times 0.35}$
Heat loss/°C of structural components	=	Surface area (m^2) x 'U' value (W/m^2 °C)
Total structural heat loss (W)	=	(Internal temperature − External temperature) x structural heat loss per °C
Ventilation heat loss (W) at minimum ventilation rate	=	0.35 x minimum ventilation rate (m^3/hr) x difference in temperature between inside and outside (°C)
Total heat loss (W)	=	Structural heat loss + ventilation heat loss
Supplementary heat required (kW)	=	Heat loss through structure + Heat loss through ventilation − Heat produced by pigs and ancillary equipment.

N.B. m^3/hr x 0.59 = ft^3/min
 m^2 x 10.76 = ft^2
 Watts x 3.412 = Btu/h

110 *Pig Environment Problems*

In effect, the calculations take account of the need of pigs to feel comfortable in warm or cold weather and the fact that they and the building and its components lose varying amounts of heat. Figure 6.1 provides an insight into the mathematics behind calculating the minimum ventilation rate.

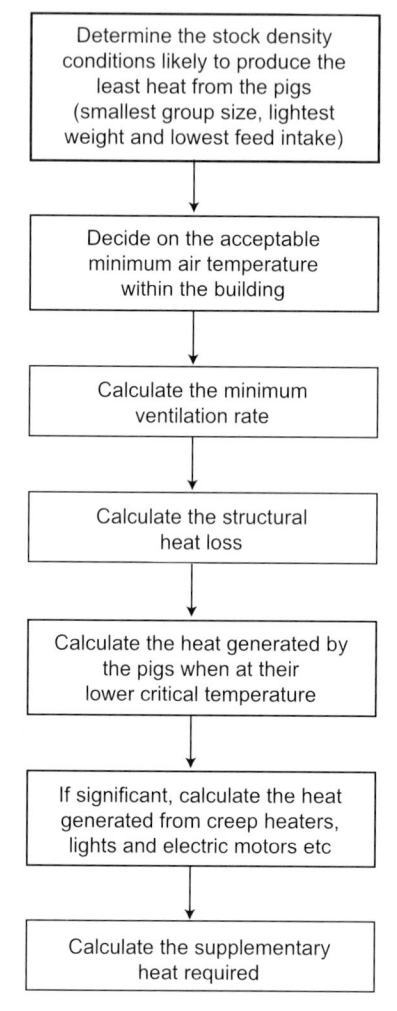

```
┌─────────────────────────────────┐
│ Determine the stock density     │
│ conditions likely to produce the│
│ least heat from the pigs        │
│ (smallest group size, lightest  │
│ weight and lowest feed intake)  │
└─────────────────────────────────┘
                │
                ▼
┌─────────────────────────────────┐
│ Decide on the acceptable        │
│ minimum air temperature         │
│ within the building             │
└─────────────────────────────────┘
                │
                ▼
┌─────────────────────────────────┐
│ Calculate the minimum           │
│ ventilation rate                │
└─────────────────────────────────┘
                │
                ▼
┌─────────────────────────────────┐
│ Calculate the structural        │
│ heat loss                       │
└─────────────────────────────────┘
                │
                ▼
┌─────────────────────────────────┐
│ Calculate the heat generated by │
│ the pigs when at their          │
│ lower critical temperature      │
└─────────────────────────────────┘
                │
                ▼
┌─────────────────────────────────┐
│ If significant, calculate the heat│
│ generated from creep heaters,   │
│ lights and electric motors etc  │
└─────────────────────────────────┘
                │
                ▼
┌─────────────────────────────────┐
│ Calculate the supplementary     │
│ heat required                   │
└─────────────────────────────────┘
```

Figure 6.1 Calculation of the minimum ventilation rate and any requirement for extra heat input.

Figure 6.2 outlines the various stages involved in the calculation of the maximum ventilation rate. On pig units where all-in/all-out production is practised rather than a continuous throughput system, the maximum ventilation rate is likely to be much higher. This arises because on a continuous throughput system the average maximum weight in the building is a key factor since pigs of a wide weight range share the building. This fundamental difference is overlooked on many pig units. Furthermore, pig farmers who elect to increase slaughter weights must not just take account of the increased floor area required, but also pay particular attention to the reality that the maximum ventilation rate must be higher for an unchanged number of heavier pigs in a fixed area.

```
┌─────────────────────────────┐
│   Determine the maximum     │
│   number of pigs and their  │
│   total maximum liveweight  │
└─────────────────────────────┘
              │
              ▼
┌─────────────────────────────┐
│  Calculate the heat output at│
│  their upper critical temperature│
└─────────────────────────────┘
              │
              ▼
┌─────────────────────────────┐
│    Calculate the maximum    │
│  ventilation rate to keep the room│
│   temperature down to no more│
│    than 3°C above the hottest│
│      ambient temperature    │
└─────────────────────────────┘
```

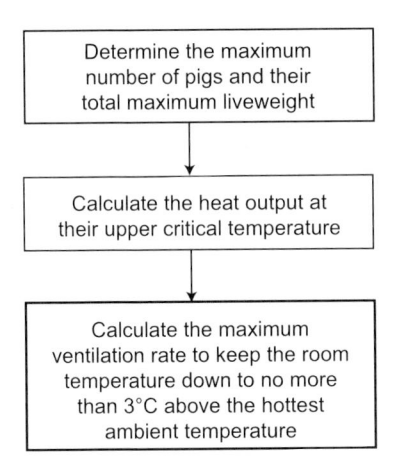

Figure 6.2 Calculation of the maximum ventilation rate needed for hot conditions.

The theory and practice of minimum ventilation rate

The minimum ventilation rate has to cater for the least number of pigs, at their lightest weight and at their lowest feed intake. Heat loss from the building, heat generated by the pigs and ancillary equipment plus any supplementary heat are also taken into account.

During very cold weather, particularly when stocking density and body weights are low, air-flow rate is reduced to a level which:

- provides an acceptable air quality,
- does not allow the build-up of carbon dioxide,
- prevents condensation,
- allows good air mixing, and
- does not make the pigs feel cold because of draughts at pig level.

112 Pig Environment Problems

These factors help define the *minimum ventilation rate*. It is concerned with the *level of dilution* of air within a building rather than the rate of removal of it. The *rate of dilution* determines the concentration of airborne contaminants in the room. When ventilation rate is increased, contaminants within the pig's environment are diluted more. Each unit volume of air leaving the room, therefore, carries less contaminants and so it follows that their concentration within the confined air space will be reduced. A practical difficulty is that the pig farmer has no accurate measure of the rate of contaminant production; scientists, as yet, have not been able to define precisely what should be regarded as an acceptable level of contamination. Figures currently used merely reflect a consensus of scientific opinion. (Table 6.2)

Table 6.2 Guidelines for minimum ventilation rates for pigs.

	m^3/h	
Weaned sows	13	(could be increased if on ad-lib feed)
Pregnant sows	13	
Farrowing sows	Around farrowing 13	Near weaning 39
Weaners (5-15kg) (11-33lb)	On arrival 1.3	By departure 3.8
Growers 15-35kg) (33-77lb)	3.8	6.6
Finishers 35-100kg (77-221lb)	6.6	11.3

N.B. $m^3/h \times 0.59 = ft^3/min$

On the farm, the pig keeper's nose tends to be the main measure of contaminants. It is not the most reliable sensor the human nose fails to smell carbon monoxide or carbon dioxide at all and many people are unable to differentiate, e.g. between 10ppm and 30ppm ammonia. Inevitably, pig keepers become accustomed to the environments in which they work and rarely have the opportunity to test air quality on other pig units. It is little wonder that there are problems getting the aerial environment right in well stocked modern pig buildings.

Practical Considerations 113

Often the minimum ventilation rate is increased because the stockperson believes that the noxious gases within the pig building are too concentrated. If the ammonia level were, e.g. 25ppm, then to reduce it to 20ppm, the ventilation rate would need increasing by 20%. This might involve increasing the minimum fan speed from 500 m³/hr (295ft³/min) to 600 m³/hr (354ft³/min). Typically this could be achieved by speeding up a 450 mm (18 inch) diameter fan from the 10% to the 12% setting. This would result in, e.g. five air changes per hour, i.e. a complete air turnover every twelve minutes. At what would still be a low ventilation rate setting, it would actually take about an hour to get the ammonia level down to the new level of 20ppm. When people compare this to the "instant" response when a domestic appliance is adjusted, this seems like a long time. Hence, because of enthusiasm or impatience, many people tend to set the minimum ventilation rate too high so that they can quickly observe the impact of the change. Under the circumstances described, it would be commonplace for the ventilation rate to have been doubled. The practical reality is that pig buildings tend to operate at more than the recommended ventilation rate. Another practical problem is that minimum ventilation rate recommendations are a compromise and assume an average pig. In any pen of pigs there is likely to be a wide range of weights and the viability of individuals will vary accordingly within the group; porcine physiology is not concerned with averages.

Table 6.3 indicates that the sensible heat output of a lightweight (5kg/ 11lb) pig at weaning is only around 19 watts. This value assumes that, even with an underweight such as this, the pig is actually eating, since food generates heat output. In practice many pigs within the weaning group, particularly the lighter ones, will be temporarily stressed and will not be eating, hence, their contribution to sensible heat output would be lost. Many pigs, therefore, require a minimum ventilation rate setting at lower than theoretical recommendations. In this way the specific needs of individual pigs competing within a group would more likely be met.

Throughout many hours of the day and many days of the year, pig ventilation systems work towards the bottom end of their range. However, when the occasion demands it, much higher ventilation rates have to be brought into play to cope with high temperatures. Despite this hot weather requirement, it is vital that the ventilation system must work well at other times when the ventilation rate is a low percentage of total capacity. In other words, modern pig housing systems demand the provision of a

114 Pig Environment Problems

Table 6.3 Sensible heat loss of pigs at various production stages.

	Heat produced in watts	
Weight (kg)	At lower critical temperature	At upper critical temperature
1	7	4
5 (Before weaning)	30	22
5 (At weaning)	19	11
10	47	35
20	76	57
40	122	94
60	156	121
80	184	144
100	206	163
Dry sow	189	142
Lactating sow	319	272

N.B. 1 kg = 0.4536 lb
      1 watt = 3.412 Btu/hr

wide range of ventilation. The system must, over a wide range of circumstances, maintain good air distribution and even temperatures, this is sometimes known as the *competence* of the ventilation system. This phenomenon is particularly important in housing systems where there is a wide weight-range within the same building, e.g. in weaning to slaughter environments.

Checking room temperature regularly helps ensure that pig buildings are adequately ventilated (Airflow Developments Ltd).

Basic design decisions often compromise the competence of a ventilation system. For example, a speed regulated fan cannot achieve a ventilation rate providing less air movement than 10% of the maximum capacity. The practical reason for this is that below 10% speed, the fan develops such little pressure so that often it is overridden by the wind, i.e. the actual output becomes unreliable. Operating within a range that gives between 100% and 10% of ventilation capacity, i.e. a ratio of 10:1 is adequate in most growing pig circumstances. However, in batch smoked nursery accommodation, this is not so. Under such circumstances it is best to use two or more fans to improve the discrimination of control at low ventilation rates. For example, in a small nursery room which has a maximum ventilation requirement that could be provided by one large fan, it is better to use two small fans to give better control at low ventilation. If, initially one of these fans is engaged at 10% of its own capacity, this equates with 5% of the maximum ventilation rate for the whole room. Another strategy is to use a combination of speed control and time switching, i.e. a fan at 10% 'working half-time' provides only 5% of total capacity for the house.

In practice the minimum ventilation rate is set to remove the main contaminants produced by the pigs: carbon dioxide and moisture. As pigs grow, they eat more and respire more and so produce more carbon dioxide, moisture and energy. The biochemistry driving this is explained in the following equation:

$$C_6H_{12}O_6 + 6O_2 \longrightarrow 6CO_2 + 6H_2O + \text{ENERGY}$$
$$\text{Carbohydrate} \quad \text{Oxygen} \quad \text{Carbon dioxide} \quad \text{Water}$$

Since more carbon dioxide and water vapour is produced as the pigs grow, the minimum ventilation rate is often increased in order to remove them. However, the increased energy output results in a rise in room temperature. This in turn results in the ventilation controller increasing the ventilation rate to keep the temperature in the room stable. Hence, increasing the minimum setting based on assumed gaseous contaminant production only, whilst ignoring the increased energy output of the pigs, can result in over ventilating at the minimum rate. Effective ventilation, inevitably, has to be a compromise between applying established science and getting the job done in a changing complex and demanding environment.

116   *Pig Environment Problems*

Ventilation not only dilutes aerial contaminants, it also creates airflow patterns inside pig buildings. These airflow patterns will vary according to factors such as season of the year, whether pigs are active or resting and the stocking density. Within compact fan-ventilated insulated buildings, the speed and direction of the in-coming air has the major bearing on airflow pattern. Air movement in large, naturally ventilated buildings is more influenced by wind speed, wind direction and siting of the building. The location of air inlets and outlets are also key factors. Provision of a compact kennel within the larger building helps reduce the impact of excessive air movement on the pigs.

When the ventilation rate is increased to dilute the concentration of gaseous contaminants, it also removes heat. If the sensible heat output from the pigs within the room is insufficient to achieve the target temperature, then supplementary heat has to be provided at additional cost. The commercial reality is such that money is first spent providing this heat and then more is spent getting rid of it along with the various aerial contaminants.

Money is spent providing heat and then more is spent getting rid of it (Gerry R Brent)

Pig buildings should, therefore, be carefully designed to a specification whereby structural heat loss and uncontrolled ventilation heat loss (through leakage) are minimised. In any given circumstances a "balancing point" is reached whereby total heat loss equates with the heat produced by the pigs and any supplementary heat added.

Tables 6.4 and 6.5 refer to heat losses in well insulated pig nursery and finishing rooms, where pigs have been provided with minimum floor areas compliant with the UK Pig Welfare Code.

Table 6.4 The impact of room size on structural heat losses in weaner housing at different ambient temperatures. (source: Farmex)

130 pig room	Room sized for 15kg (33lb) At 12°C (54°F) ambient	At −5°C (21°F) ambient	Room sized for 30kg (66lb) At 12°C (54°F) ambient	At −5°C (21°F) ambient
	Structural heat loss			
Structural heat loss	900W	1750W	1200W	2500W
	Ventilation heat loss			
@3% of capacity	650W	1300W	1300W	2650W
@5% of capacity	1050W	2200W	2100W	4400W
@10% of capacity	2100W	4400W	4200W	8800W

N.B. 1 watt = 3.412 Btu/hr

Table 6.5 The impact of room size on structural heat losses in finishing housing at different ambient temperatures. (source: Farmex)

130 pig room	Room sized for 50kg (110lb) At 12°C (54°F) ambient	At −5°C (21°F) ambient	Room sized for 75kg (165lb) At 12°C (54°F) ambient	At −5°C (21°F) ambient
	Structural heat loss			
Structural heat loss	950W	2550W	1200W	3250W
	Ventilation heat loss			
@3% of capacity	1000W	2700W	1500W	4050W
@5% of capacity	1650W	4500W	2500W	6750W
@10% of capacity	3350W	9000W	5000W	13,500W

N.B. 1 watt = 3.412 Btu/hr

In both instances there are 130 pig places, one room accommodates pigs to 15kg (33lb) and the other to 30kg (66lb). The tables provide examples of structural and ventilation heat losses in the two rooms. More structural heat is lost when the ambient temperature is lower (-5°C v 12°C) or (21°F v 54°F) and more is lost when the room is bigger. The trend, driven by the need to reduce both pig stress and labour input, is to move pigs less and this could have implications for ventilation efficiency. There is a danger in larger rooms accommodating a wider age and size range of pigs that the effective minimum ventilation rate could be too high because of the overall greater ventilation capacity required in that building and the increased risk of leakage. This will inevitably lead to increased supplementary heating costs. Whilst Tables 6.4 and 6.5 illustrate the impact of room size on structural heat loss, it is important to understand that ventilation heat loss is much greater. This means that careful attention needs to be paid to ventilation control. Since the aerial contaminants, carbon dioxide and water vapour are related to pig size rather than room size, this could be an argument in favour of raising the minimum ventilation rate in wide weight range pig buildings. However, in larger buildings, air leakage alone often results in the minimum ventilation rate being exceeded. Operators should always remember that ventilation control systems respond to room temperature which is dominated by the results of the heat output of growing pigs. A room or building of healthy, feeding and growing pigs is the surest way to have a well ventilated building.

Keeping pigs comfortable under changing conditions

Pigs will only thrive when they feel comfortable. Housing pigs in an environment where room temperature is below the Lower Critical Temperature could compromise the welfare of the pig and would not be commercially advantageous. When room temperature approaches the Upper Critical Temperature (UCT), pigs become stressed and unproductive. Commercial pig producers strive to keep pigs within their comfort zone and, provided all other aspects of the environment are acceptable, a temperature within the pig's lying area which is just above the LCT is desirable. Given the extreme variation between summer and winter temperatures and diurnal variation, air-flow within pig buildings is deliberately varied in an attempt to achieve the target temperature. Stocking density is also a key factor, plus the availability of supplementary heat or additional cooling measures at the limits of the comfort zone.

Practical Considerations 119

As room temperatures increase, ventilation rates can be increased so that the temperature remains within a selected range. This generally, but not necessarily, is dependent on the use of automation which ensures that air throughput is regulated appropriately. It could well involve increasing air movement by running fans at a higher speed, engaging more fans or increasing the inlet area, particularly in naturally ventilated systems.

Pigs give out body heat and this waste heat increases room temperature. A pen of 21 bacon pigs together could have to dissipate up to 3kw (10,236 Btu/h) of heat from their bodies.

Table 6.3 indicates how the degree of body heat loss from pigs varies according to their liveweight. It also shows that this heat loss, known as the *sensible heat loss* is minimised when pigs feel cold whilst it is maximised under warmer conditions.

Despite the increased heat loss at higher temperatures on particularly hot days when stocking density is high, there is the possibility that the Upper Critical Temperature will be reached. In the absence of air conditioning or radical changes to air movement patterns, the ventilation system, therefore, eventually fails to keep the pig within its comfort zone. Most pig buildings are designed so that, at the maximum ventilation rate, room temperature can at the best only be kept down to 3 or 4°C above that of the incoming air. In effect, there is an environmental compromise.

Table 6.6 provides guidelines for maximum ventilation rates.

Table 6.6 Guidelines for maximum ventilation rates for pigs (m³/h per pig place).

	Internal temperature rise above ambient	
	3°C (37.4°F)	*4°C (39.2°F)*
Weaned sows	209	156
Pregnant sows	134	100
Lactating sows	545	406
Weaners (15kg/33lb)	44	33
Growers (35kg/77lb)	84	63
Finishers (100kg/221lb)	145	108

N.B. k³/h x 0.59 = ft³/min

It highlights the need to increase ventilation as body weight increases.

120 *Pig Environment Problems*

The data also illustrate the need to increase airflow rate as feed intake increases during lactation and also the need to boost the minimum ventilation rate when pigs gain body weight.

Figure 6.3 attempts to bring together the key elements involved in the effective ventilation of pigs.

Coping with very hot weather

Varying the speed of air movement over pigs in hot weather can markedly enhance pig comfort and improve pig behaviour. However, ventilation has its limits. In hot zones of the world, sprinkler systems and misting are frequently used in conjunction with ventilation systems to relieve heat stress in pigs. A sequence of warm summers, the need to use labour more efficiently, welfare issues and greater responsibility for due diligence has increased interest in the potential of sprinkler systems in temperate zones.

In Denmark, during summer the temperature in pig buildings often exceeds 22°C (72°F) for up to 60 days per year. Finishing pigs in particular start to wallow when room temperature exceeds 22°C (72°F) and, in particular, this can lead to behavioural problems and vices, especially on totally slatted systems where wallowing is not generally an option.

On totally slatted systems, wallowing is not an option (PIGSPEC).

Practical Considerations 121

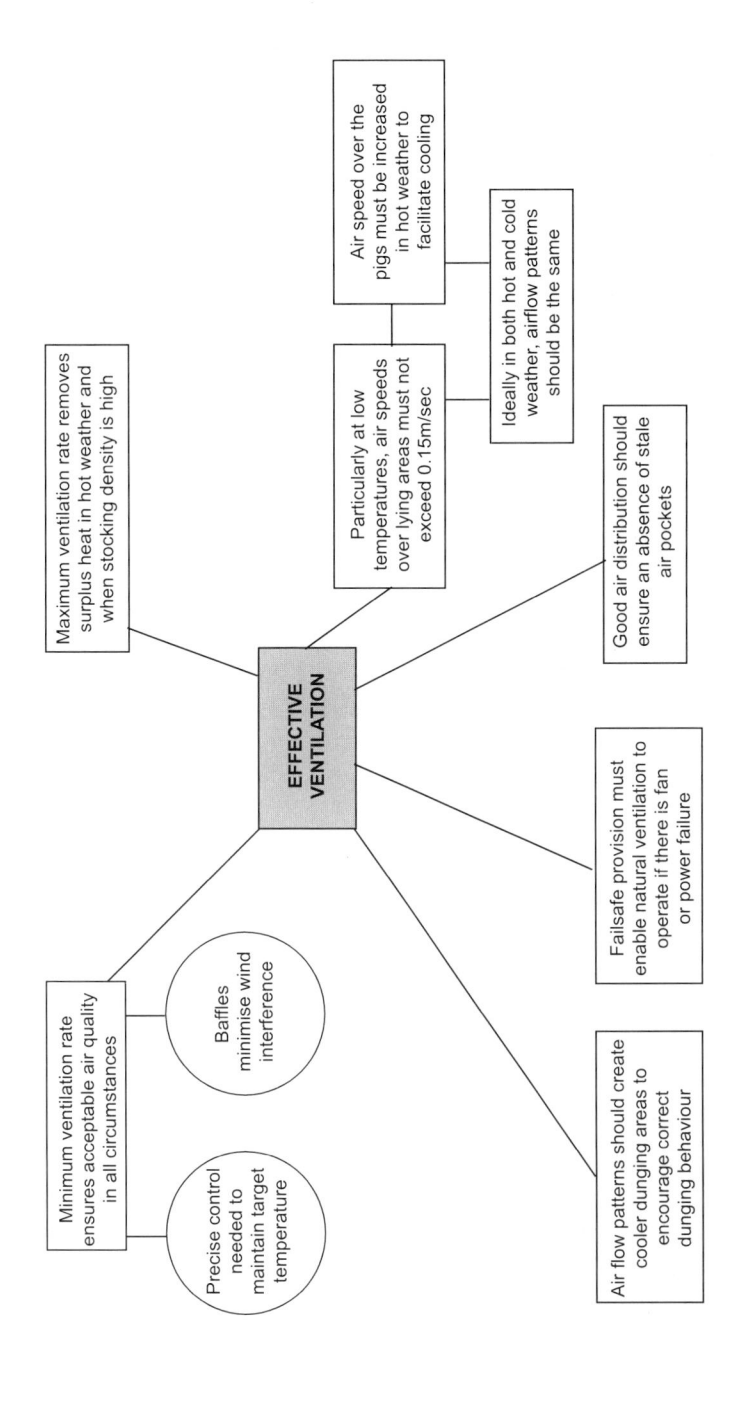

Figure 6.3 A blueprint for better ventilation.

- EFFECTIVE VENTILATION
  - Maximum ventilation rate removes surplus heat in hot weather and when stocking density is high
  - Air speed over the pigs must be increased in hot weather to facilitate cooling
  - Particularly at low temperatures, air speeds over lying areas must not exceed 0.15m/sec
  - Ideally in both hot and cold weather, airflow patterns should be the same
  - Good air distribution should ensure an absence of stale air pockets
  - Failsafe provision must enable natural ventilation to operate if there is fan or power failure
  - Air flow patterns should create cooler dunging areas to encourage correct dunging behaviour
  - Minimum ventilation rate ensures acceptable air quality in all circumstances
  - Baffles minimise wind interference
  - Precise control needed to maintain target temperature

122   *Pig Environment Problems*

Pedersen *et al* (1998) studied the impact that a sprinkler cooling system could have on aggression and pen cleanliness of finishing pigs during hot weather in Denmark. Trials were undertaken on three different farms:

Farm 1   Ad-lib, dry fed/fully slatted
Farm 2   Wet fed in troughs (restricted)/fully slatted
Farm 3   Ad-lib, dry fed/partially slatted

A misting system for hot weather cooling (Farmex).

Auxillary ventilation system from a polythene duct working in conjunction with a misting system (Farmex).

Practical Considerations 123

The trials took place between June and September and on each farm both control and sprinkler systems were evaluated. Sprinkling took place between 8am and 9pm, twice per hour when room temperature exceeded 22°C (72°F). The sprinklers were mounted over the dunging areas and set at 35 litres/hour (7.7 imp. gals/hr or 9.3 U.S. gals/hr), each sprinkling session lasted for just 30 seconds.

Figure 6.4 summarises the overall results and indicates that, across the three farms, pigs remained significantly cleaner if they were subject to sprinkling with water (P < 0.001).

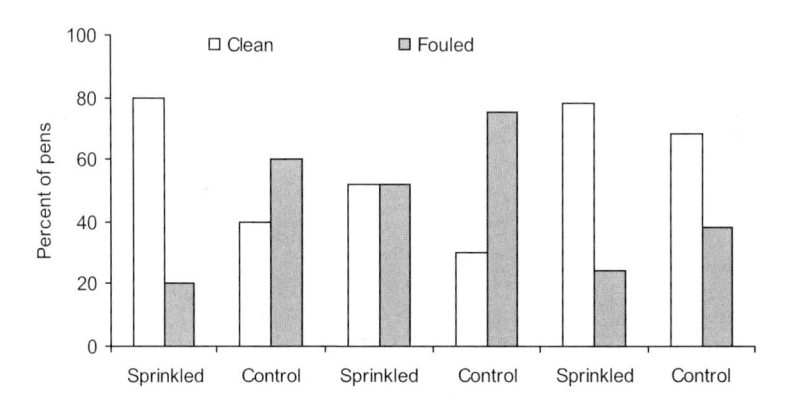

Figure 6.4 Effect of sprinkling on pen hygiene.

Results shown in Table 6.7 indicated a significant reduction in aggression in the three herds in which, as it happened, aggression was not particularly a problem.

Table 6.7 Effects of hot weather sprinkling on finishing pigs (Pedersen *et al*, 1998).

	Total aggression %					
	Farm 1		Farm 2		Farm 3	
	Sprinkler	Control	Sprinkler	Control	Sprinkler	Control
	52.3[a]	36.9[b]	60.1	63.6	66.3[a]	35.4[b]

	Time spent lying %					
	Farm 1		Farm 2		Farm 3	
	Sprinkler	Control	Sprinkler	Control	Sprinkler	Control
In dunging area	25.9[a]	33.0[b]	21.9	26.4	20.5[a]	35.3[b]
In lying area	74.1[a]	67.0[b]	78.1	75.4	79.5[a]	64.7[b]

[a, b] denote statistical significance at P<0.05

The results also indicated that lying behaviour was significantly better when sprinklers were used. Hence, water-sprinkling systems can go a long way in helping to reduce inherent shortcomings in the dependence on ventilation systems during hot weather. A common practical problem with this system is a failure to regulate the water flow correctly and also there is a tendency to use the device when air temperatures do not justify it. Sprinkler systems, however, offer the additional facility of enabling pig buildings to be automatically and thoroughly pre-soaked prior to power-washing operations. Some pig unit managers report good results where pig operated drinkers which incorporate a sprinkler facility are located in the dunging area.

7

USING AIR MOVEMENT TO IMPROVE PIG ENVIRONMENT

Principles of air movement

Air movement patterns have a major influence on pig comfort and behaviour. Hence those involved in both the design and management of the pig environment must have a grasp of the basics of air movement. Pig keepers will be able to respond better to the changing needs of the pig if they understand why air moves and how it moves.

Think of air as an invisible mass of matter. Rather like a vehicle free-wheeling, it will move if it has momentum. Air also moves as a result of pressure difference, i.e. in an attempt to equilbrate, it will move from an area of high pressure into an area of lower pressure. Air which has a different density or buoyancy than the mass of air around it will also move, e.g. dense, cold air will fall and light warm air will rise. When a mass of dense, cold air enters a warm pig building it tends to fall but, once it has been warmed-up by the body heat of the resident pigs, it attempts to rise. The best ventilation systems exploit these basic laws of physics in such a way that the comfort of the pigs is not impaired whilst the air quality is safeguarded.

Pigs and pig buildings interrupt natural airflow patterns. Air movement from outside a pig building to inside the building and from inside a pig building to the outside is governed by several factors. A "stack effect" arises because of pressure differences between warm inside air and cold outside air. Wind blowing across the external structure of the building brings about pressure differences resulting in air movement. Fans also cause air to move by physically exaggerating pressure differences between the inside and outside of a building.

Figure 7.1 based on work undertaken at Scotland's Centre for Rural Building and The National Institute of Agricultural Engineering, provides a guide to likely pressure changes exerted by a strong breeze on a compact high pitch (22.5°) pig building.

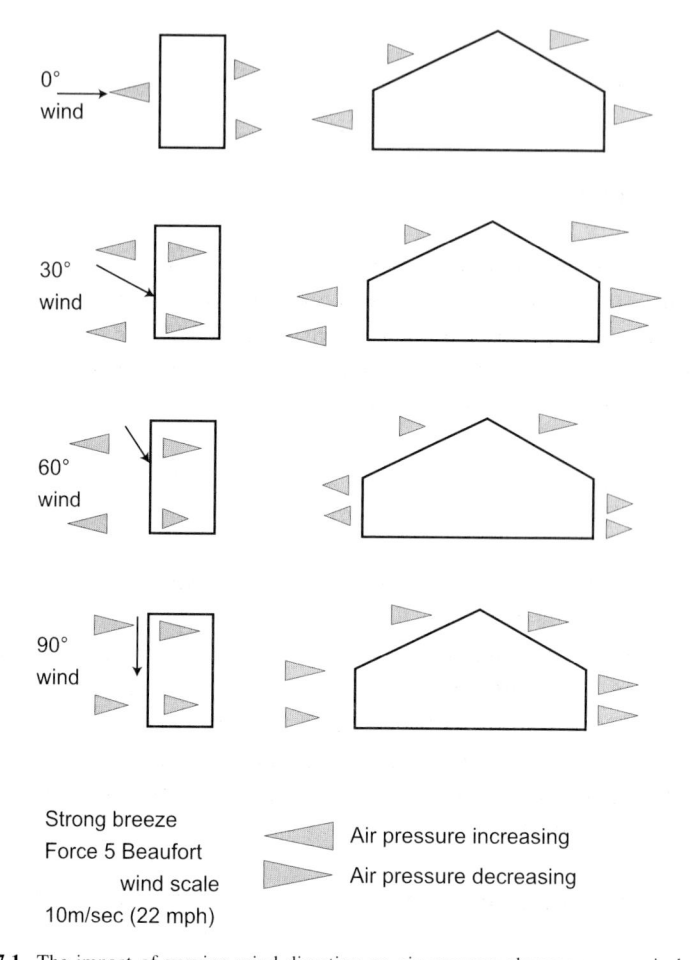

Figure 7.1 The impact of varying wind direction on air pressure changes near a pig building.

Typically this might measure 29m (95ft 2ins) long, 9m (29ft 6ins) wide and 2.4m, (7ft 10ins) to the eaves. The diagram indicates that the windward side of a building is under positive pressure. However, by the time the air mass reaches the leeward side, there is a change from positive to negative pressure. If the wind strikes the building at 0°, pressure at the ridge and the leeward side would be equal on a steep pitch building such as that illustrated. (Wells and Moran, 1979).

Exposure to wind and shelter from it must be taken into consideration when siting a new pig building and its impact on existing buildings must also be taken into account. Abel-Rehiem and Douglas (1976) undertook research on the impact of close-siting of agricultural buildings.

Figure 7.2 outlines their research findings which indicated that unless adjoining buildings were sited at least double the ridge height apart, all the surfaces on the leeward side of the building would be under negative pressure. Given the opportunity, new buildings should always be sited in a freestanding situation. In very exposed situations windbreaks such as plastic webbing, trees or bushes can help reduce the impact of wind on a building. Earth banks, bumps and natural ground contour also influence air movement.

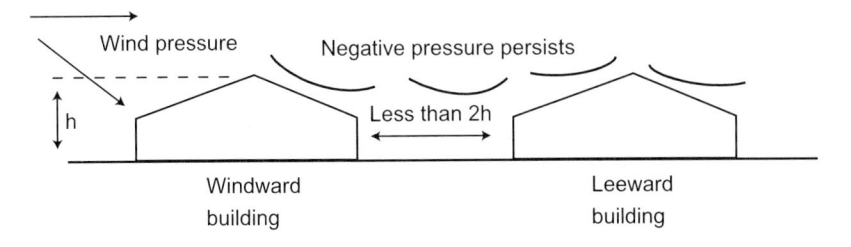

Figure 7.2 The impact of siting two building too close.

Pressure differences around buildings are inevitable and will result in some airflow through the building structure because of air leaks. Furthermore, airflow through inlets, outlets and fan shafts also result in pressure differences. During Force 5-8 (10-19m/sec), i.e. powerful 22-43 mph winds, air speed would be sufficient to override installed fans which will generally resist a back pressure of 25N/m² (0.1 inch W.G.); back pressure arises because pressure is lost when air has to pass through openings at inlets and outlets.

An appreciation of the mathematical relationship between air speed, rate of air removal and inlet area is a great help in understanding how ventilation works and why, sometimes, problems arise. Figure 7.3 comprises three simple mathematical equations which demonstrate the inter-dependence of air speed, fan capacity and inlet area. If fan power throughput remains constant and air inlet area is decreased, air speed will increase. The mass of incoming air, in effect, assumes the shape of the inlet facility.

Air speed m/sec (ft/min)	=	Rate of removal m³/sec (ft³/min) Inlet area m² (ins²)	Inlet ↑ Air speed ↓
Inlet area m² (ins²)	=	Rate of removal m³/sec (ft³/min) Air speec m/sec (ft/min)	Inlet ↓ Air speed ↑
Rate of air removal m³/sec (ft³/min)	=	Air speed m³/sec (cu. ft/min) × inlet area m² (ins²)	Air speed/ inlet ↑ Rate of air removal ↑

Figure 7.3 The relationship between air speeds, rate of air removal and inlet area.

The theory of stack effect

Since warm air inside a building is less dense than cold air outside, it is more buoyant and so tends to rise and, if there is provision, it will leave the building through the roof or a high point in the gables. As this warm air escapes at a high level, more cold air replaces it entering at a lower level and for example can be drawn into the building through eaves inlets.

Figure 7.4 depicts the key components of stack effect, i.e. pressure and temperature gradients between outside air and inside air and the difference in height (the stack) between the point where incoming air meets warm air already in the building and where warm air meets the outside cold air as it leaves the building.

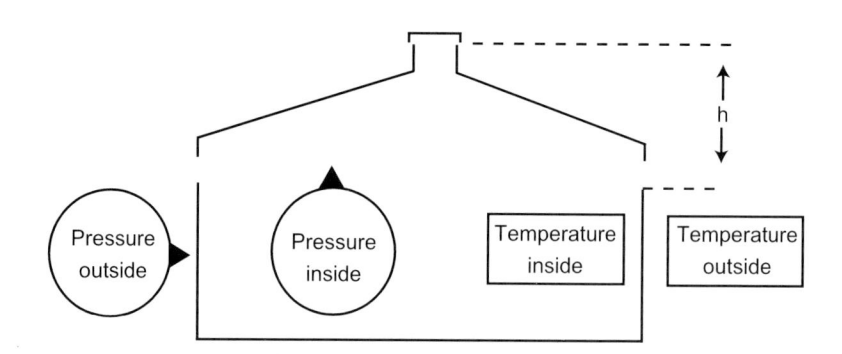

Figure 7.4 Ventilation resulting from stack effect.

Bruce (1973) has defined the mathematical relationships involved in this physical process.

Pressure difference between outside and inside of a building = 0.041 H (ti − to) N/m²

When: ti = house temperature (°C)
to = outside temperature (°C)
H = difference in height between outlet and inlet (m)

Table 7.1 gives examples of the degree of stack effect attainable for a range of temperature lifts and given distances between air inlet and outlet. Previously it has been mentioned that fans will usually resist back pressure of 25N/m² (0.1 inch W.G.) and this can be overridden by a strong wind blowing between 22 and 43 mph.

Table 7.1 Pressure difference between the outside and inside of a building arising from Stack-Effect (N/m²)

Temperature difference		5°C (41°F)	15°C (59°F)	25°C (72°F)	30°C (86°F)
Height difference (m)	(ft)				
0.5	1ft 8in	0.10	0.31	0.51	0.62
1.0	3ft 3in	0.21	0.62	1.03	1.24
3.0	9ft 9in	0.62	1.85	3.08	3.69
5.0	16ft 3in	1.03	3.08	5.13	6.15

NB:N/m² x 0.0043 = ins. Water Guage (WG)

Referring to Table 7.1, it can be seen that pressure differences due to stack effect are very low compared to the full power of fans and strong winds.

Key practical points are:

- The substantial pressure difference at 5°C (41°f) between stack heights of 1.0m and 5.0m (0.21 N/m² v 1.03N/m²).
- At the same two stack heights there is a big increase in stack effect when the inside/outside temperature lift is as much as 30°C (86°f) (1.03 N/m² v 6.15 N/m²).

130 Pig Environment Problems

Put another way, in a compact farrowing house when there is only a small temperature differential between inside and outside, there would be little movement of air due to stack effect. However, in a large portal-frame structure, e.g. with a ridge height at 6.0metres (19ft 8 ins), particularly during extreme differences in inside and outside temperature, there would be much more potential for stack effect ventilation.

The significance of inlets and outlets

In livestock ventilation systems, as the name suggests, the "inlet" is where the ventilating air comes into the airspace and the "outlet" is where the contaminated air leaves the airspace. A ventilation fan could be an inlet or an outlet.

Air moves as a result of its momentum and its response to a pressure change or because of density differences with the surrounding air. Carpenter (1973) undertook trials which helped explain the different impact that inlets and outlets have on air movement patterns. Airflow patterns were studied as air might enter a building through a 620 mm (24 ins) diameter hole. When air speed through the opening was 10 metres/second (2000 ft/min), even at a distance of 10 metres (33ft) inside the building, its speed was still as high as 3.9 metres/second (780 ft/min), i.e. it acted as a jet of air (Figure 7.5).

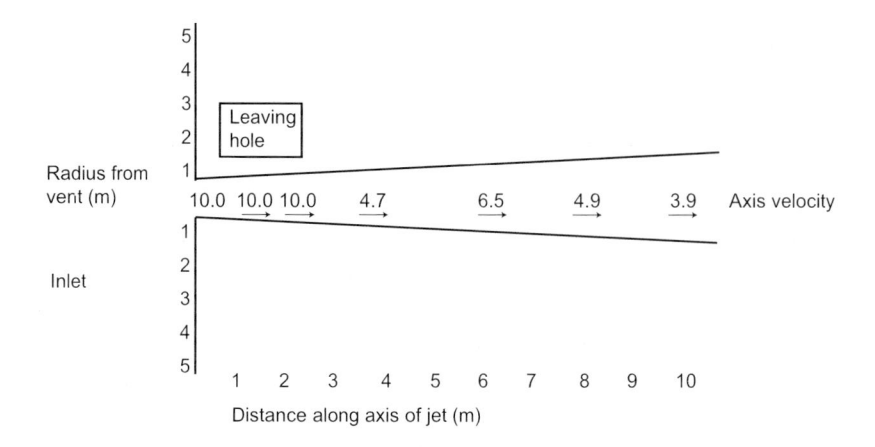

Figure 7.5 Speed of air leaving a 620mm (24 inch) hole at 10m/sec (2000ft/min). (source: Carpenter (1973))

Using Air Movement to Improve Pig Environment 131

Carpenter's studies indicated that as air leaves through a 620 mm (24 inches) diameter opening, the air movement characteristics change markedly. Within the building momentum is lost and is barely measurable 10 metres (33ft) from the outlet. When pressure differences occur and are detected by the air stream, air moves radially towards the opening, gathering momentum as it travels. Even so, when the air is as near as 0.5 metres (20 inches) from the outlet opening, its speed is only 0.5 metres/second (100 ft/min) (Figure 7.6).

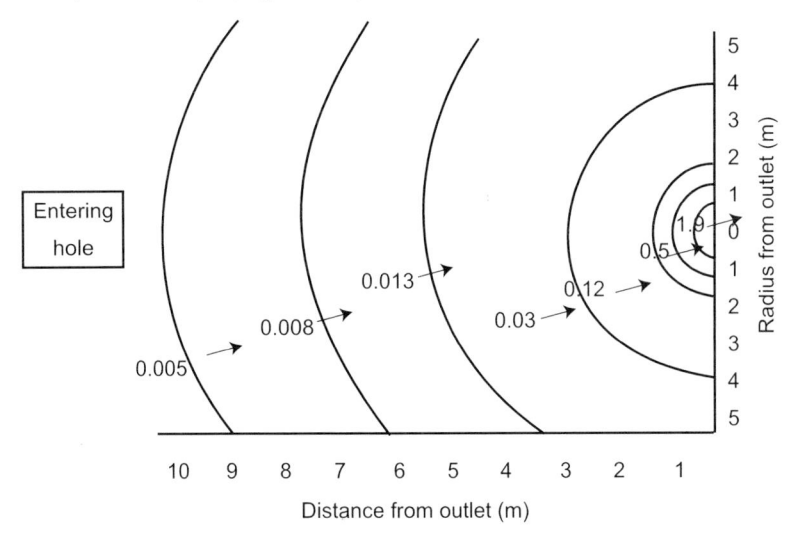

Figure 7.6 Speed of air entering a 620mm (2 ft hole), air speed at hole being 10m/sec (2000ft/min). (source: Carpenter (1973))

This phenomenon is often regarded as the first law of ventilation, i.e. *within a building inlets rather than outlets control air movement patterns.* Provided outlets are protected by, e.g. back-draught shutters to prevent wind interference, there is no reason why they cannot be located close to the pigs.

When air first enters a building it is not only subject to horizontal forces because of its momentum but also subject to vertical forces because of density differences, together these forces greatly influence where the air will flow and how quickly it will flow there.

Usually, when air enters a building it is colder and denser than the air already within it. Consequently, incoming air tends to drop and, particularly in cold weather, this can be disastrous for pigs lying near the inlet, unless there is effective baffling. At the best, it will lead to dirty pigs and a deterioration in growth rate and feed conversion efficiency, but it is also highly likely to lead to the onset of respiratory disease. Figure 7.7 illustrates a conventional inlet both without and with an inward baffle. Since the inlet gap is generous, air speed is reduced such that an inward speed of one metre/second (200ft/min) would be typical. Even so, without internal baffling there is a tendency for the incoming air to fall on entry to the building.

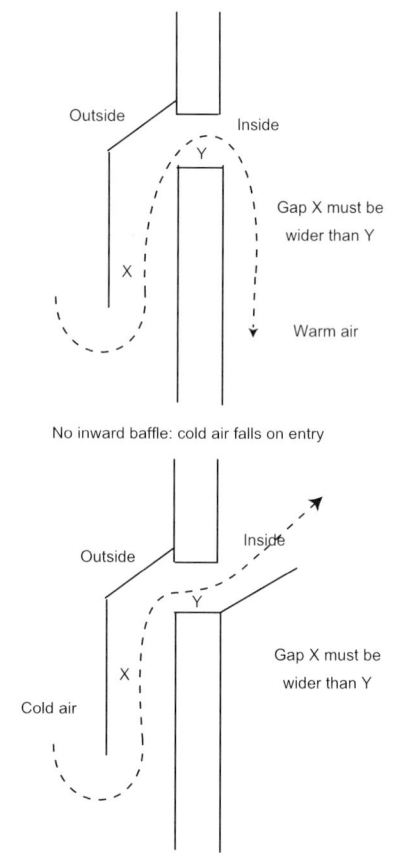

Figure 7.7 Baffled air inlets.

Figure 7.8 depicts different temperature scenarios for incoming air through a conventional wide opening baffled inlet. Whereas these extremes reflect climate changes throughout the year, sometimes, diurnal variations in

Using Air Movement to Improve Pig Environment 133

weather also have this effect. Hence, automation of the inlet facility is a good idea.

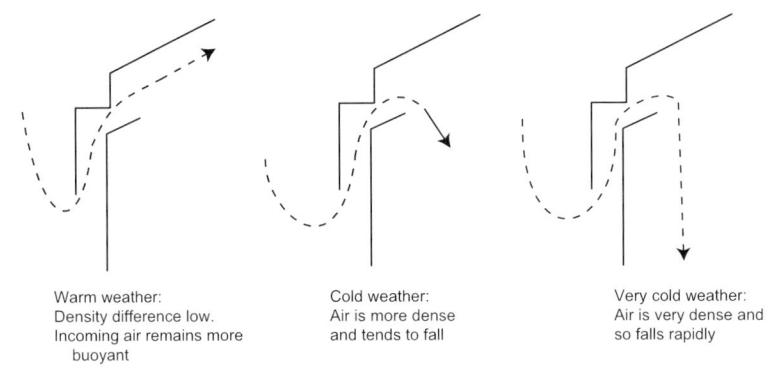

Warm weather:
Density difference low.
Incoming air remains more buoyant

Cold weather:
Air is more dense and tends to fall

Very cold weather:
Air is very dense and so falls rapidly

Figure 7.8 The effect of temperature differences on an air jet through a conventional hopper inlet.

Open-topped wind-protected air inlet (Farmex).

Wind protected baffled air inlets showing open top sealed base and extractor fans mounted below (Farmex).

The importance of inlet design has long been understood. The maintenance of a constant and relatively high inlet velocity at the inlet will ensure the incoming fresh air is evenly distributed and thoroughly mixed within the airspace. The volume of air through the space is varied to control temperature.

Volume = Area x Speed

In the above equation, since the *speed* is roughly constant and the *volume* is being varied, it follows that the *area* (size of the inlet) must also be varied. Clearly, this is best done automatically. Control systems today typically match the volume of ventilating air (controlled by sequential on/off switching of propeller fans) to careful regulation of the inlet area in order to provide a constant, predictable and mixed airflow.

Factory-made automatically controlled inlets are now manufactured. Typically they comprise durable, self-supporting moulded urethane box-like structures. Their dimensions are such that they can be mounted directly into brick or block walls. 'Slimline' versions are also available but these tend to be factory-fitted by specialist pig building companies who manufacture walls from compact insulated panels. Figure 7.9 shows the scope for automatically varying the degree of air inlet ranging from a narrow gap upwardly directing air during cold weather to a much wider gap which allows air to gush in horizontally during hot, still conditions.

Passively operated air inlets also enable incoming air to be varied without human interference, these resemble a 'letter-box' type structure usually made in fibre-glass or hard plastic (Figure 7.10).

The degree of inlet depends on the pressure difference resulting from the ventilation speed of fans usually mounted in side walls, the roof or gables. During cold weather they upwardly deflect air by means of a pivoted flap, but during hot weather a higher positioned pivoted deflector plate can be used to send incoming air directly over the pigs to cool them.

Air moving through a narrow gap behaves quite differently. As a jet of air flows through an air space, since this is not a vacuum, it will be subject to friction. The degree of friction can be reduced by allowing the air jet to 'adhere' to a smooth ceiling line. This tendency for a jet of air to cling to a smooth ceiling is known as the 'Coanda effect'.

Air supply for different opening angles and negative pressure (SjF test No. 82)

Air direction for different opening angles (SjF test No. 82)

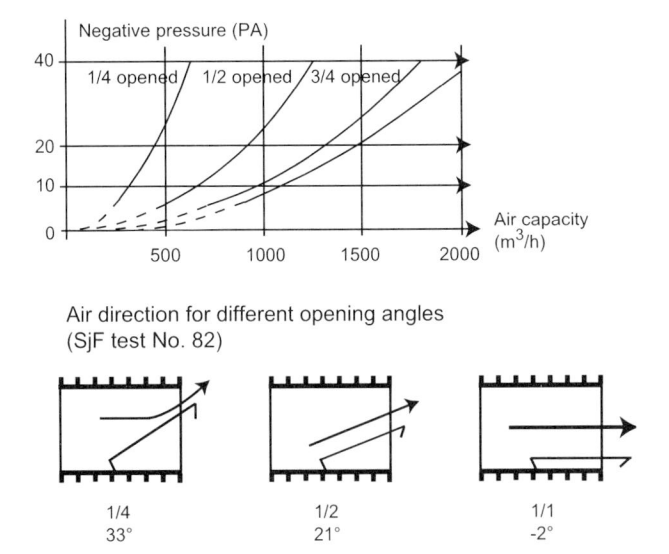

1/4	1/2	1/1
33°	21°	-2°

Figure 7.9 Automatic variable air inlet. (source: Big Dutchman)

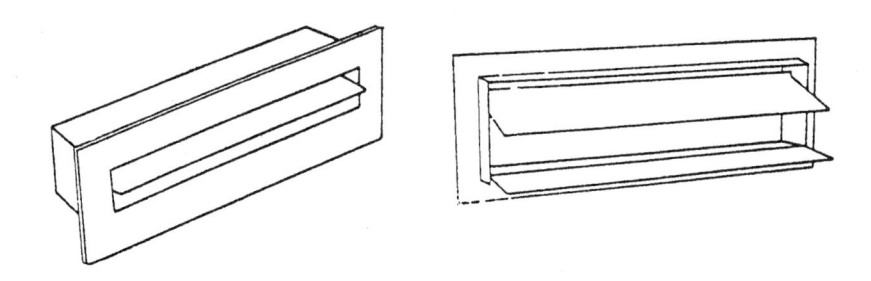

Figure 7.10 Letter-box type passive air inlet. (Farmex)

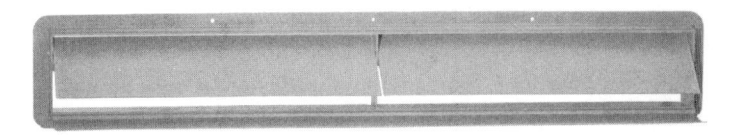

Passively operated air inlets enable incoming air to be varied without human interference. (Weiss)

136 *Pig Environment Problems*

Figure 7.11 illustrates the principle and the problems encountered when the smooth ceiling line is obstructed. The Coanda effect is commercially exploited by most high inlet speed ventilation systems. Ceiling obstructions such as battens, fluorescent lights and electrical trunking play havoc with these ventilation systems, particularly when located at right angles to the air stream and therefore need to be avoided.

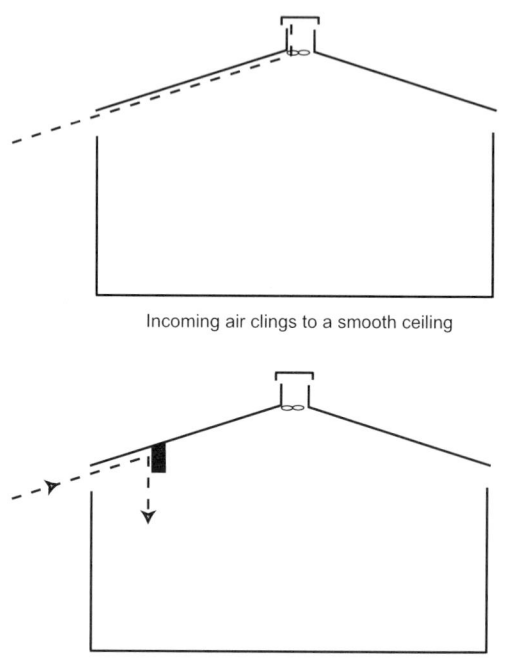

Incoming air clings to a smooth ceiling

An obstruction on the ceiling downwardly deflects incoming air

Figure 7.11 The Coanda effect.

An alternative variant of high-speed jet ventilation is the downdraught jet

Figure 7.12 shows the system which is designed for air to enter rapidly at 5 metres/second (1000ft/minute).

Its immediate downward deflection creates draughts in the dunging area running along the outside walls of the building. By the time this originally cold air has crossed the slats or dung passage, it has become warmer because of pig body heat within the building. Hence, it travels across the lying area at between 0.1 and 0.2 metres/second (100-200 ft/minute) which provides comfortable conditions and encourages pigs to remain recumbent and keep clean within the designated lying area.

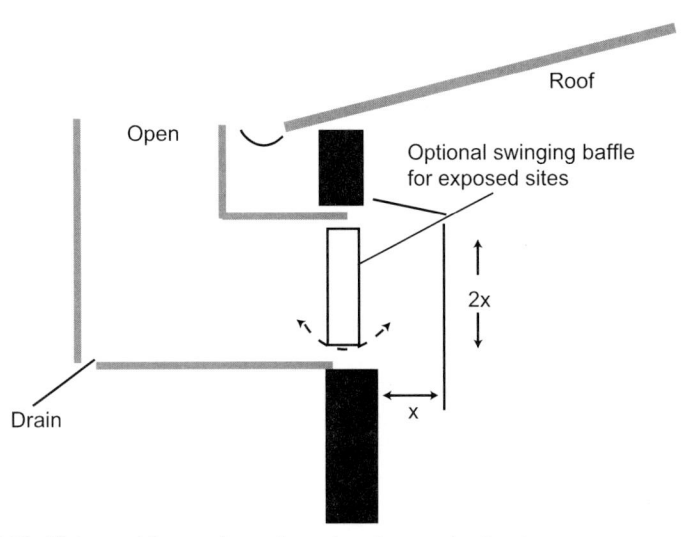

Using Air Movement to Improve Pig Environment 137

Figure 7.12 High speed jet creating a down-draught over the dunging area.

Predicting air movement patterns

When the layout of a pig building is first considered, a knowledge of air movement patterns and the ability to predict them is essential. This was made possible following some far-reaching research undertaken at Silsoe Research Institute, (Randall, 1975). Figures 7.13-7.16 illustrate different air movement patterns actually traced within a building. The diagrams represent what is likely to happen during different environmental situations with various ventilation systems.

Figure 7.13 indicates that when air is introduced horizontally at eaves height on a cold day or into a warm building on a warm day, different air movement patterns arise before exhaust air leaves through the fan outlet. The situation illustrated on the left hand side of Figure 7.13, where air enters at medium speed, results in the incoming air clinging to the ceiling line before falling. However, in the same building, slow moving air falls rapidly as soon as it enters the building. The right hand side of figure 7.13 refers to the same building operating on a warm day. After testing a range of air speeds and ventilation systems under different weather conditions, Randall was able to conclude:

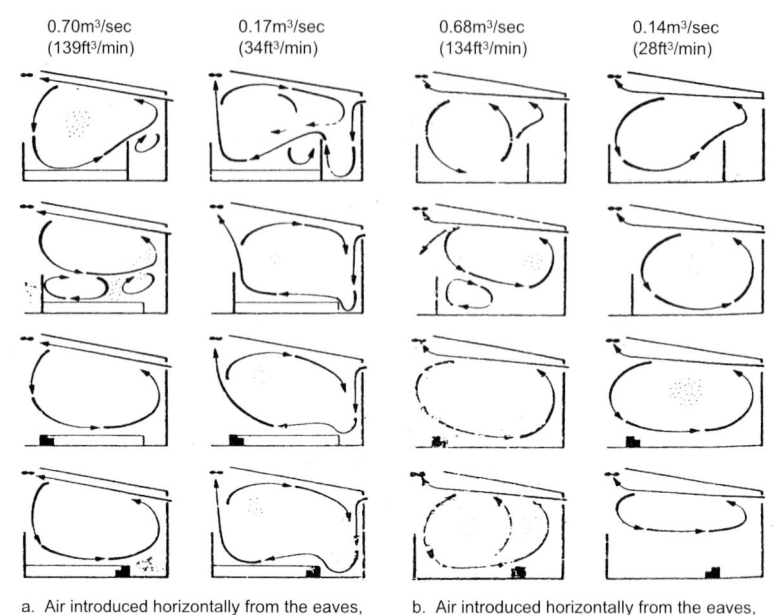

138 *Pig Environment Problems*

a. Air introduced horizontally from the eaves, non-isothermal, e.g. cold outside/warm inside

b. Air introduced horizontally from the eaves, isothermal, e.g. warm outside/warm inside

Figure 7.13 Eaves inlet ventilation system - horizontal discharge.

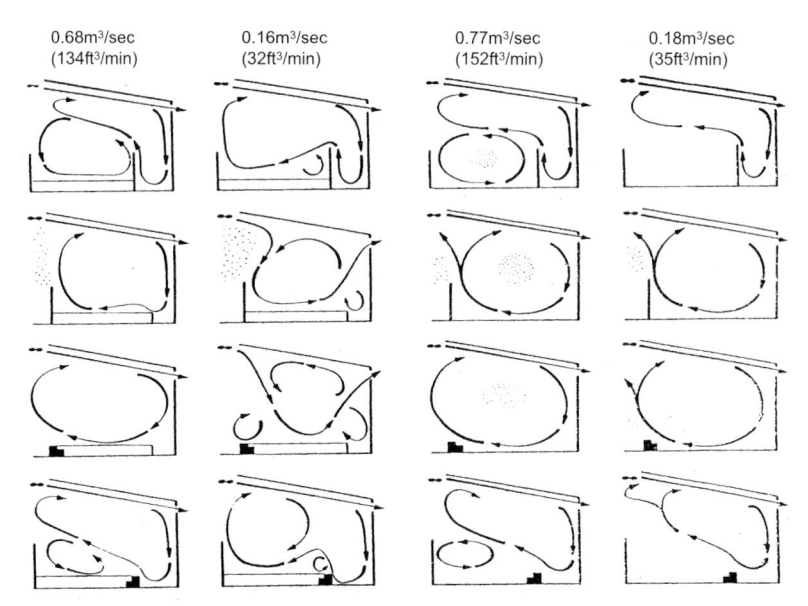

a. Air introduced horizontally from the ridge, non-isothermal, e.g. cold outside/warm inside

b. Air introduced horizontally from the ridge, isothermal, e.g. warm outside/warm inside

Figure 7.14 Ridge inlet ventilation system - horizontal discharge.

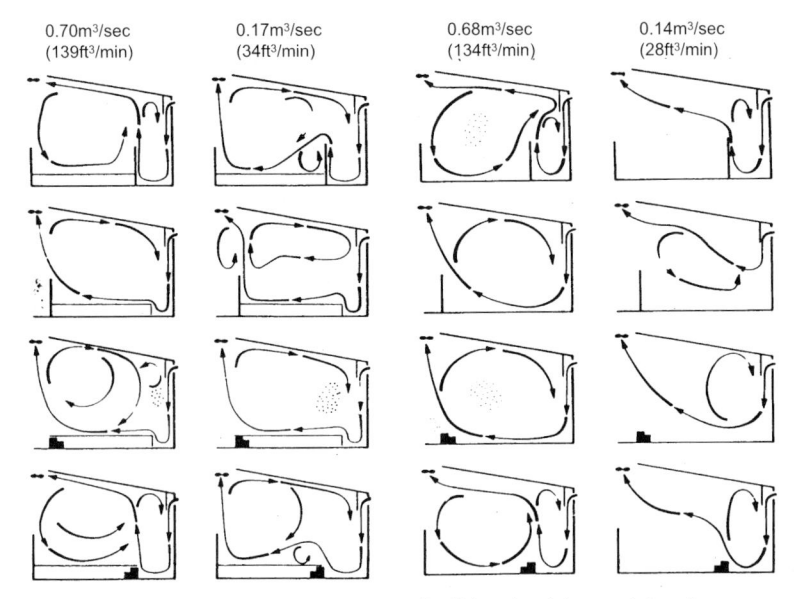

Using Air Movement to Improve Pig Environment 139

a. Air introduced downwards from the eaves, non-isothermal, e.g. cold outside/warm inside

b. Air introduced downwards from the eaves, isothermal, e.g. warm outside/warm inside

Figure 7.15 Eaves inlet ventilation system - downward discharge.

a. Air introduced downwards from the ridge, non-isothermal, e.g. cold outside/warm inside

b. Air introduced downwards from the ridge, isothermal, e.g. warm outside/warm inside

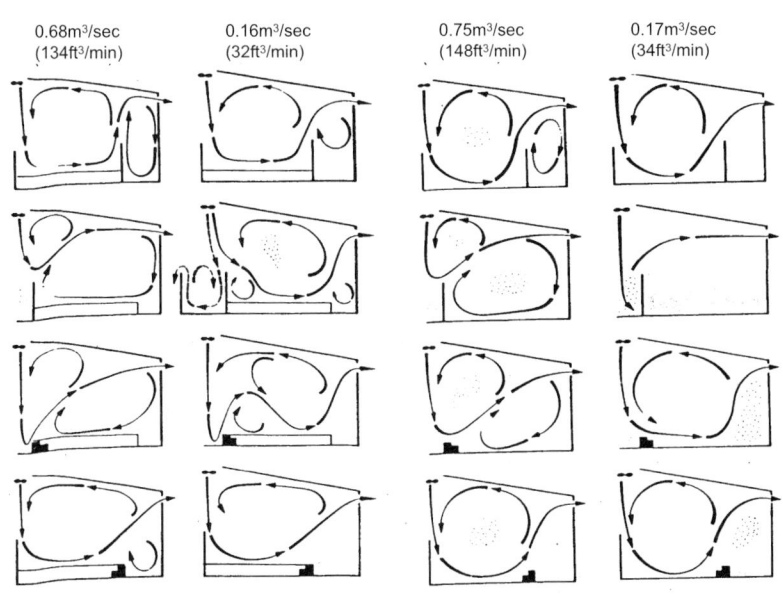

Figure 7.16 Ridge inlet ventilation system - downward discharge.

- Air in buildings always moves in rolls.
- Unless incoming air is put through a diffusing medium, the inlet facility controls the shape of the roll of the air.
- Unless the speed of the incoming air is kept very high, eg. 5metres/sec (1000 ft/minute), air behaves differently depending on the variation between the outside and inside temperature.
- Obstructions on the roofline and solid pen divisions interfere with airflow patterns.

The difficulties of efficiently ventilating pig buildings over a range of conditions reflect that iso-thermal and non-isothermal conditions not only occur in the different seasons of the year, but also arise within the same 24-hour period. This situation is commonplace, particularly during spring and autumn. As a result of the associated changes in air movement patterns, respiratory stress is greatest under those conditions.

Propeller fans

Propeller fans are commonly used in pig buildings to control the rate of air movement. These fans are able to move large volumes of air at reasonable capital and low running costs. Since the pressure development of the propeller fan is limited, this must be allowed for when the ventilation system is designed. Table 7.2 provides an approximate guide to maximum fan capacity over a range of commonly used fan sizes at a typical back pressure.

Table 7.2 Guide to typical maximum air volume throughput with commonly used propeller fans.

Fan size – impeller diameter		Volume throughput	
mm	in.	m^3/hr	ft^3/min
315	12	1500	885
350	14	2500	1475
400	16	3500	2065
450	18	5000	2950
500	20	7500	4425
630	25	10000	5900
710	28	12500	7375

Using Air Movement to Improve Pig Environment 141

The exact performance will vary according to the manufacturer's specification. On large-scale enterprises, particularly in the USA, tunnel ventilated finishing systems are used and often the fan diameter is 1000-2000mm (39-78 in) on these pig units. There is much scope for misunderstandings when installing ventilation systems and maximum efficiency is often not achieved. Consequently most propeller fans used in livestock ventilation are now delivered as a complete unit including mounting plate and bell-mouth.

Tunnel ventilation finishing barns coping with hot weather in America (Farmex).

Fan detail of tunnel ventilated finishing barn in America (Farmex).

In some instances a fan introducing air and another one expelling it are mounted within the same factory-made shaft and there is a linked recirculation facility. Control equipment ensures that the supply and exhaust fans operate at the same speed and this ensures that pressure balance is maintained in the building. At low outside temperatures, fresh incoming air is mixed with air already in the room so that it is warmed while being discharged into the room. The result is an absence of downdraughts and good air mixing. In very cold weather in inadequately insulated buildings with low stocking density systems capable of providing very low ventilation rates, for example those that use recirculation, end up maintaining set temperature by operating for prolonged periods at low ventilation rates thus sometimes there is a tendency to over-use the re-circulation facility and this can impair air quality.

Ventilation system design and installation requires good communication between pig farmer, design engineer, builder and electrician. If the biological needs of the pig are misunderstood or are ignored, results can be commercially disastrous

Fan ventilated systems

Broadly speaking, there are two types of fan-powered systems used to ventilate pigs:- "Pressurised" systems are those where the fan is used to blow air into the air space. "Extract" systems are those where the fan is used to blow air out of the air space. Some designs favour so called "neutral pressure" systems whereby the volume which is blown into the air space is matched by an equal volume being blown out. Globally, extract systems are widespread with sequentially controlled propeller fans combined with regulated inlet components.

Alternative or emergency ventilation

Concern about the welfare of intensively housed pigs has resulted in regulations in many countries which help ensure, for example, that alternative provisions of ventilation are made at the building design stage to operate in the event of system or power failure. Priority should be given to the signalling of any fault so that qualified staff can respond quickly and effectively. The normal functioning of the main ventilation system should not be compromised by alternative ventilation components. However, where response times are likely to exceed 15 minutes, then automatic system response must be considered.

Using Air Movement to Improve Pig Environment

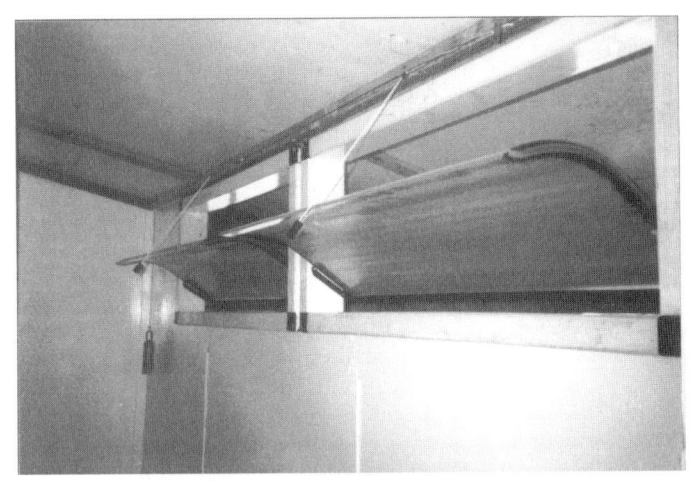

"Optiflex" cable controlled automatic air inlets (Farmex).

Provision of a fail-safe device is vital for coping with emergencies (Peter Heath).

The Welfare of Farmed Animals (England) Regulations 2000 (S.I. 2000 No. 1870), Schedule 1, paragraphs 18-21, state that:

18   All automated or mechanical equipment essential for the health and well being of the animals shall be inspected at least once a day to check there is no defect in it.

144   *Pig Environment Problems*

"Cool cell" air inlet in a tunnel ventilated gestation barn in America (Farmex).

19   Where defects in automated or mechanical equipment of the type referred to in the paragraph above are discovered, these shall be rectified immediately, or if this is impossible, appropriate steps shall be taken to safeguard the health and well-being of the animals pending the rectification of such defects including the use of alternative methods of feeding and watering and methods of providing and maintaining a satisfactory environment.

20   Where the health and well-being of the animals is dependent on an artificial ventilation system:-
   a.   provision shall be made for an appropriate back-up system to guarantee sufficient air renewal to preserve the health and well-being of the animals in the event of failure of the system; and
   b.   an alarm system (which will operate even if the principal electricity supply to it has failed) shall be provided to give warning of any failure of the system.

21   The back-up system shall be thoroughly inspected and the alarm system shall each be tested at least once every seven days in order to check that there is no defect in the system and, if any defect is found (whether when the system is inspected or tested in accordance with this paragraph or at any other time) it shall be rectified immediately.

The regulations also state:

61. All mains electrical equipment should meet relevant standards and be properly earthed, safeguarded from rodents and out of the pigs' reach.
62. All equipment, including feed hoppers, drinkers, ventilation equipment, heating and lighting units, fire extinguishers and alarm systems, should be cleaned and inspected regularly and kept in good working order.
63. All automatic equipment used in intensive systems should be thoroughly inspected by the stock-keeper, or other competent person, not less than once each day to check that there are no defects. Any defect must be rectified immediately.

Future trends in ventilation

In the past, ventilation systems for pig buildings have been greatly influenced by the need to eliminate noxious gases from buildings. The emphasis has been on providing a better air quality for housed pigs and those that attend them. Consequently, discharge of dust and foul air into the atmosphere has been widespread with little regard for the environment beyond the pig building.

The impact of the Integrated Pollution Prevention and Control regulations in Europe and the sentiment behind them is likely to have a marked influence on design requirements in future pig building ventilation systems. In the UK, trials are currently underway at Silsoe Research Institute which aim to improve the design of slatted floor buildings and so reduce the potential for ammonia emissions. It is already acknowledged that ammonia emissions are less from part-slatted buildings. Hence, there is likely to be a trend towards the use of part-slatted systems and this also happens to comply with future welfare recommendations. The need to eliminate high levels of noxious gases through ventilations systems will, in the future, be reduced as a result of new thinking in the collection and storage of pig slurry. Slurry systems will be designed such that there will be less surface area of the slurry pit in contact with air. There is likely to be more use of non-fouling convex plates under the slats, less storage of slurry within the same air space as the pigs and more storage of slurry in lidded tanks beyond the pig buildings.

146 *Pig Environment Problems*

In Holland a ventilation system known as 'The MTC Stable' is giving good results, especially in pig finishing systems. The essentials of the system are outlined in Figure 7.17.

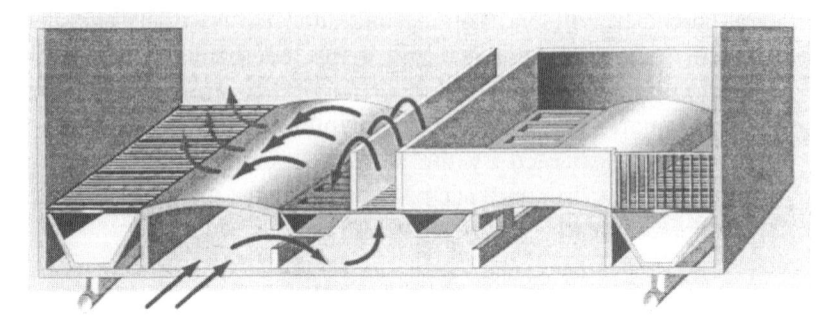

Figure 7.17 The 'MTC Stable' novel ventilation system.

The concept involves bringing draught-free air into the building much nearer to pig level than in traditionally ventilated systems. If need be, this incoming air can be water spray-cooled in summer and it is quite possible to reduce air temperature at pig level to 4°C (39°F) below that of the incoming air from outside the building. Air is introduced from under the central access passage. It rises and emerges near the solid pen front to which it clings before gently falling into the lying area. Air then follows the lines of the convex floor before it is extracted above the shallow slurry accumulation. Results indicate less build-up of noxious gases on account of the design of the slurry channel and there are indications that the pigs are more comfortable on account of the availability of 'fresh' air.

Kennels and naturally ventilated systems

Most fan ventilated systems aim for the precise control of air throughput within an insulated shell, the creation of a draught-free lying area and a draughty dunging area. Conversely, naturally ventilated kennel or bungalow systems involve the creation of a compact, draught-free, preferably bedded, lying area located within a vast air space where air movement is just tempered rather than controlled. Traditionally, kennel systems have been built within airy large un-insulated structures with eaves height 3m (10ft) or more. Space boarding, i.e. vertically placed timbers usually measuring about 125mm (5in) wide, 20 mm (0.75in) thick with

20 mm (0.75in) air gaps between are fixed above external block walls and the gables. Depending on siting and wind direction, some air leaves the building via the gables and side walls but ideally goes out through the ventilated ridge. The most unsophisticated systems lack any winter draught-proofing over the necessarily wide kennel entrances and there is little more than a space-boarding knock-out facility for coping with very still days and high temperatures. The cornerstone of the system is the provision of a warm, dry, draught-free bedded lying area and the facility for the pig to have freedom to move within the colder, draughtier and wetter environment beyond the kennel. Particularly when kennels are not insulated, there is a need to minimise the cubic capacity of the kennel in order to achieve the required temperature lift within it. This presents accessibility and observation difficulties if the pig keepers are required to enter the kennel and a predisposition for dust to interfere with the respiratory system of the pig. Kennels about 1.5m, (5ft) high or more help ensure better observation by the pig keeper but run the risk of becoming cold and draughty, particularly if the kennel tops are higher than the last course of blockwork on the side walls (Figure 7.18).

Naturally ventilated portal-frame building with bedded kennels (JSR Genetics/BHR Communications).

Especially during windy weather, down draughts through open ridges can be a problem in naturally ventilated buildings.

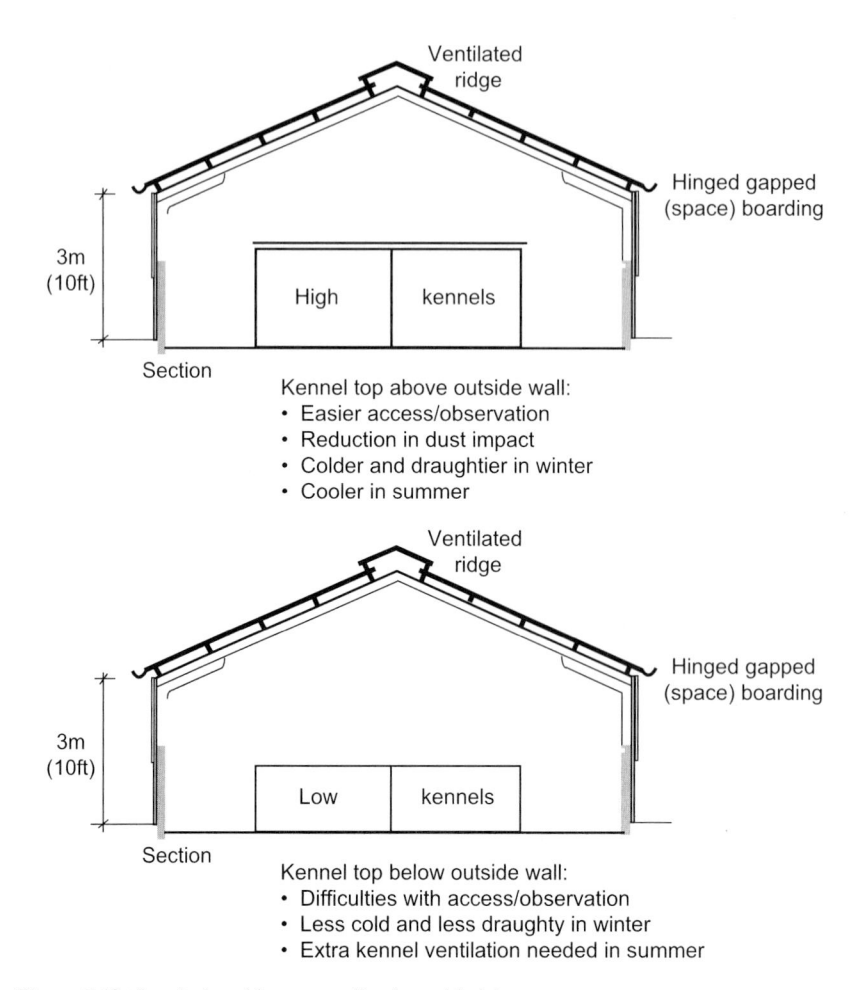

Figure 7.18 Practical problems regarding kennel height.

Turbulence on the gable ends results in cold air falling rapidly along the mid-line of the building. The impact of this can be reduced by blocking the ridge opening immediately above the first third of the end pens (Figure 7.19). Netting of open ridges is also important to prevent bird invasion.

Lack of control of air movement is a fundamental problem with space-boarded, naturally ventilated buildings. Particularly during wintry weather, cold air entering at a low level can be a problem since this predisposes the pigs to respiratory stress. Many farmers react to this practically by covering the lower third of the space-boarding with plastic sheeting or plywood cladding just for the winter period.

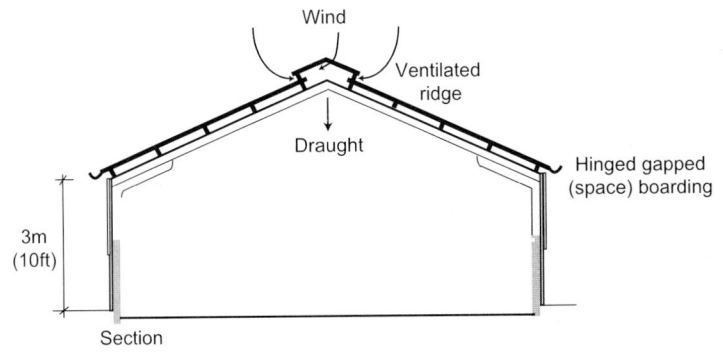

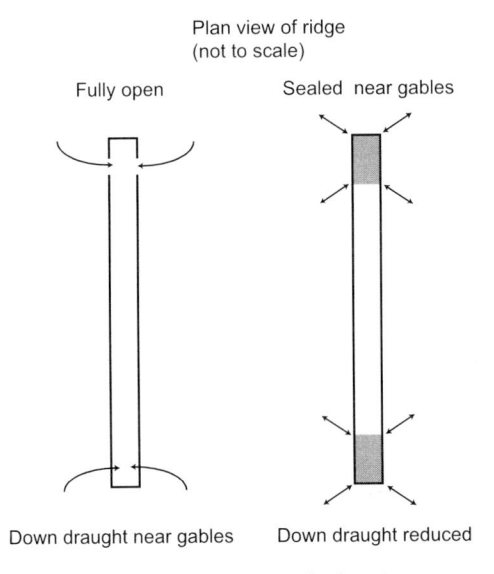

Using Air Movement to Improve Pig Environment 149

Figure 7.19 Down draught near gables and reduction strategy.

The traditional kennel system is obviously imprecise, however it does facilitate the generous use of straw and the production of high levels of stackable farmyard manure with a high organic matter content. It also has certain appeal on mixed farms in that the portal-frame structure is regarded as a more adaptable building with opportunities for long-term alternative use.

Some naturally ventilated portal-frame pig buildings in recent years have been subjected to more "fine-tuning". This has come about because of a better understanding of the environmental needs of the pigs and the use of new technology. Strategic use of lightweight frame-mounted perforated coated polyester material is now regarded as an alternative to space-boarding. Sometimes this perforated plastic is mechanically operated by means of a roller mechanism.

Some 'fine-tuning' for a naturally ventilation building (Galebreaker Products).

It is seen as a cost-effective and practical alternative to using heavy duty hinged space-boarding or cumbersome knock-out panels to combat "dead" areas during hot, still conditions.

Curtain-sided buildings have been popular in pig buildings in the USA for some years and this technology is being increasingly used to increase the scope for varying air throughput in traditional naturally ventilated pig buildings in the UK.

The system lends itself to automation and has the big advantage that air can be progressively allowed into the building from the eaves level downwards. Hence the likelihood of introducing draughts into the kennel area during cold, windy weather is reduced.

Automatic control of natural ventilation

An obvious shortcoming with natural ventilation is the need for manual adjustments to be made in response to changing conditions. Weather

Using Air Movement to Improve Pig Environment 151

Curtain sided buildings automatically allow air into a building from the eaves downwards. (Simco Systems)

patterns are often such that frequent adjustments to the ventilation are required and pig keepers find this tedious and not very productive. The practical outcome is that manual adjustments are often restricted to "emergency" changes during extremes of weather. Automatic Control of Natural Ventilation (ACNV) goes a long way to overcoming this practical difficulty without being burdened with significant running costs. ACNV aims to keep house temperature at the optimum level for pig production efficiency. The internal temperature of the pig building is continually monitored by a controller which then alters the ventilation rate when required typically by adjusting flaps or curtains. It can also be used to control natural ventilation of kennels

Essentially an ACNV system comprises a control unit, a sensor, adjustable flaps usually in the side walls of the building and a motor or linear actuator which opens and closes the flaps. At pre-set intervals the controller instructs the sensor to check air temperature. Depending on the reading obtained, the sensor then switches power to the opening and closing mechanism so that the ventilation rate can be adjusted. Figure 7.20 summarises how the system works and Figure 7.21 represents the cross-section of a building fitted with ACNV.

'ACNV' system on grower building in the UK (Farmex).

Centre pivoted 'ACNV' flaps with externally mounted linear actuator (Farmex).

Advantages of ACNV include:

- A low energy consuming system with low energy costs.
- Good temperature control, especially at low ventilation rates.
- Capable of operating at low air speeds inside the building, even when wind speeds outside are high.
- The system does not encourage dust movement.
- Compared to fan ventilation, ACNV is a 'silent' system.
- Provision of a lighter environment which is pleasant for pigs and people.
- Less likelihood of disasters if electricity fails.

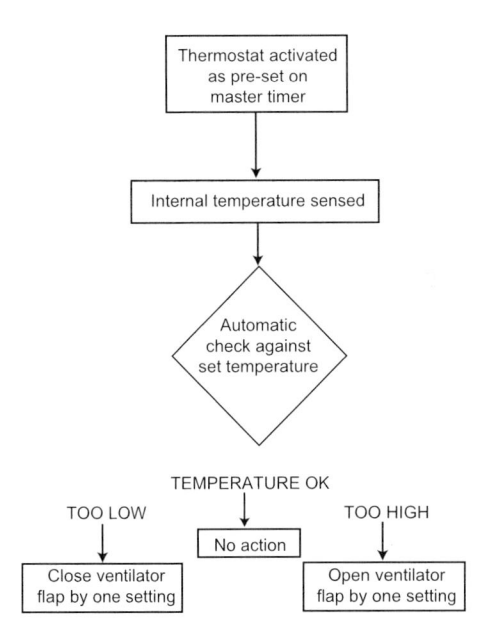

Using Air Movement to Improve Pig Environment 153

Figure 7.20 ACNV controller cycle of operation.

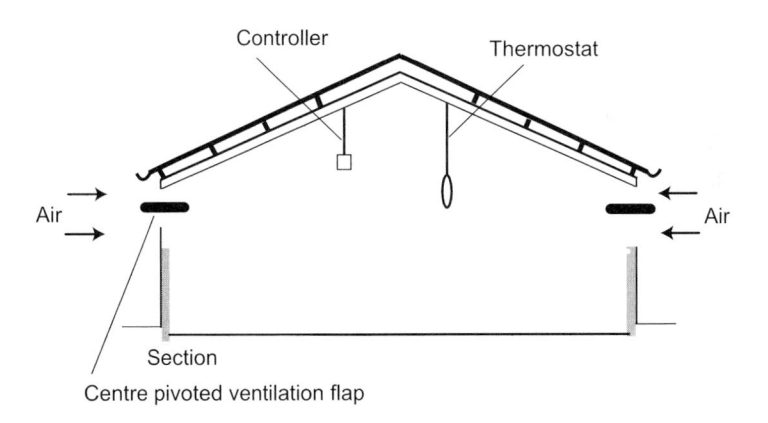

Figure 7.21 Automatic control of natural ventilation.

There are, however, some shortcomings:

- Not suitable for all sites, particularly when buildings are close together.
- Wide openings in side walls allow bird invasion unless proofing measures are undertaken.

- ACNV does not control the direction of air movement within a building.
- Bigger flaps in side walls will be required as batch production and heavier slaughter weights become more commonplace.
- More difficult to dry out the building following power-washing operations.

Designers of ACNV systems usually work on the following principles:

- An air speed of 1m/second (200ft/minute) is assumed since totally still conditions are a rarity.
- Flap sizes are based on the sensible heat output of the pigs.
- Local meteorological data is used to ascertain summer design temperature.
- These are based on historical readings over a ten-year period when the maximum ambient temperature exceeded the suggested design temperature for just 1% of the time.
- Reference tables are used which take account of the summer design temperature and, from this, flap size is calculated.

Practical experience indicates that attention to detail in the design and execution of the system is paramount. Lightweight, panel-type centre pivot flaps lend themselves to linking to a direct drive system which is subjected to minimum resistance from the flaps. On-site monitoring indicates that there is a gradual deterioration in performance over time. This can be rectified with a simple maintenance programme. ACNV is an energy efficient system which, in an era of the Climate Change Levy and Integrated Pollution Prevention and Control regulations, must be seen as an opportunity.

Pig Environment Problems

The authors have aimed to enable those working in the various sectors of the pig industry to develop a better understanding of the environmental needs of the pig and the consequences of getting it wrong. A future publication is planned which will focus on case-study work and the scope for cost-effectively improving pig environment.

REFERENCES

Abel-Rehiem, A.H.A., Douglas, M.P. (1976) Effect of adjacent structures on farm building ventilation. The Agricultural Engineer, Winter.

Baekbo, P. (1989) Air quality and pig herd health (in Danish) PhD Thesis. The Royal Veterinary and Agricultural University, Copenhagen.

Baldwin, B.A. (1969) British Veterinary Journal. **125** 281-288.

Bigelow, J.A., Houpt, T. R. (1988) Physiology and Behaviour. **43** 99-109.

Boon, C. R. (1981) The effect of departures from lower critical temperature on the group postural behaviour of pigs. Animal Production. **33** 71.

Brouns, F., Edwards, S.A. (1994) Applied Animal Behaviour Science. **39** 225-235.

Bruce, J.M. (1973) Natural ventilation by stack effect. Farm Buildings Progress. **32**.

Bruce, J.M. and Clarke, J.J. (1979) Models of heat production and critical temperature for growing pigs. Animal Production **28** 353.

Carpenter, G.A. (1973) Ventilation of buildings for intensively housed livestock. Nottingham Easter School.

Cole, D.J.A., Chadd, S.A. (1989) Voluntary feed intake of growing pigs. The Voluntary Feed Intake of Pigs, British Society of Animal Production. Occasional Publication. **13.**

Crook, B, Robertson, J.F., Travers Glass, S.A., Botheroyd, E.M., Lacey, J and Topping, M.D. (1991) Airborne dust, ammonia, micro-organisms and antigens in pig confinement houses and the respiratory health of exposed farm workers. American Industrial Hygiene Association Journal. **52** 271 - 279.

Dantzer, B.R. (1986) Journal of Animal Sciences **62** 1776-1786.

Day, J.E.L., Kyriazakis, I. and Lawrence, A.B. (1995) Applied Animal Behaviour Science. **42** 193-206.

DEFRA (2002) Guidelines: Environmental impact assessment for use of uncultivated land or semi-natural areas for intensive agricultural purposes.

DEFRA (2003) Code of Recommendations for the Welfare of Livestock: Pigs. **3.**
DEFRA (2003) Code of Recommendations for the Welfare of Livestock: Pigs.
DEFRA (2004) Integrated Pollution Prevention & Control – A Practical Guide – Edition 3.
Drummond, J.G., Curtis, S.E., Meyer, R.C., Simons, J. and Norton, H.W. (1981) Effects of atmospheric ammonia on young pigs experimentally infected with *Bordella bronchiseptica*. American Journal of Veterinary Research. **42** 463-468.
Dutch Meat Board (2003) Personal communication.
Ewbank, R. (1996) Veterinary Record. **85** 183-186.
Ewbank, R. & Meese, G.B. (1971) Animal Production. **13** 685-693.
Fraser, D. (1974) Journal of Agricultural Science, Cambridge. **82** 147-163.
Fraser, D. (1987a) Applied Animal Behaviour Science. **17** 61-68.
Fraser, D. (1987b) Canadian Journal of Animal Science. **67** 909-918.
Gonyou, H.W., Chapple, R.P. & Frank, G.R. (1992) Applied Animal Behaviour Science. **34** 291-301.
Graugvogl, A. (1958) Dr Vet. Med Dissertation, Frei University Berlin, Journal No. 215. Kiley, M (1972) Z. Tierpsychol. **31** 171-222.
Harker, R.R., Ogilvie, J.R., Morrison, W.D. & Kains, F (1994) Journal of Animal Science. **72** 1455-1460.
Heber, A. J. & Stroik, M. (1988) Influence of environmental factors on dust characteristics in swine finishing houses. Proceedings of the Third Internal Livestock Environment Symposium A.S.A.E, 1-88, 291-298.
Hsia, L.C., & Wood-Gush, D.G.M. (1984) Applied Animal Ethology. **11** 265-270.
International Commission of Agricultural Engineering (1994) Aerial Environment in Animal Housing. Working Group Report Series No. **94** 1 1-116.
Kare,M.R., Pond,W.C and Campbell. (1965) Animal Behaviour. **13** 265-269.
Kay, R.M. & Leigh, P.M. (1997) Proceedings of British Society of Animal Science **11.**
Kennedy, J.M. & Baldwin, B.A. (1972) Animal Behaviour **20** 706-718.
Kiley-Worthington, M. (1977) *Behaviour Problems of Farm Animals*, Stocksfield UK, Oriel Press.
Larsson, K. (1983) Dusting reducing method when handling concentrates.

Swedish Institute of Agricultural Engineering (JTI) Report No. **399**.
Lawrence, A.B. & Terlouw, E.M.C. (1993) Journal of Animal Science. **71** 2815-2825.
McAllister, J.S.V. (1966) Gases from dung under slats. Farm Buildings. **11** 23-24.
McBride, G. (1963) Animal Behaviour **11** 53-56.
McLaughlin, C.L., Baille, C.A., Buckholtz, L.L. & Freeman, S.K. (1983) Journal of Animal Science. **56** (6) 1287-1293.
Meese, G.B., Conner, D.J & Baldwin, B.A. (1975) Physiology and Behaviour. **15** 121-125.
Nilsson, C. (1982) Dust investigations in pig houses. Swedish University of Agricultural Science. Department of Farm Buildings, Special Report No. **149** Lund.
Oldeburg, J. (1989) Geruchs-und Ammoniak-Emissionen aus der Tierhaltung. KTBL-Schrift. **333** Darmstadt .
Pearson, C.C. (1989) Air cleaning with wet scrubbers, Farm Buildings & Engineering **6** 2.
Pedersen, B.K., Petersen, L.B., Hjelholt, K., Anderson, H.M.L., Ruby, V., and Kai, P. (1998) Growing and Finishing Pigs: Cooling reduces aggressive behaviour and pen fouling. Proceedings of the 15th IPVS Congress, Birmingham, England.
Persaud, K.C., Khaffaf, S.M., Hobb, P.J. & Sneath, R.W. (1996) Chemical Senses **21**, 495-505.
Petersen, V. (1994) Applied Animal Behaviour Science. **45** 215-224.
Petersen, V., Simonsen, H, B. & Lawson, L.G. (1995) Applied Animal Behaviour Science. **45** 215-224.
Phillips, P.A., Fraser, D. (1987) Canadian Agricultural Engineering. **29** 2. 193-195.
Randall, J.M. (1975) The prediction of airflow patterns in livestock buildings. Journal of Agricultural Engineering Research. Vol **20** No. 2.
Robertson, A.M. (1971) Effect of ventilation on the gas concentration in a part-slatted piggery. Farm Buildings R&D Studies **(1)** May 1971. 17-28. Scottish Farm Buildings Investigation Unit.
Robertson, J.F. (1989) Effect of purge ventilation on the concentration of airborne dust in pig buildings, Proceedings of the Eleventh International Congress on Agricultural Engineering (Dublin).
Robertson, J.F. (1992) The influence of aerial environment in the severity of enzootic pneumonia in pigs. In: *Environmental and energy*

aspects of livestock housing. Proceedings of the seminar of C.I.G.R. Second Technical Section, Polancia, Poland.

Rushen, J. (1984) Animal Behaviour. **32** 1059-1067.

Schiffman, S.S., Stalleley Miller, E.A., Suggs, M.S., & Graham, B.G. (1995) The effects of environmental odours emanating from commercial swine operations on the mood of nearby residents. Brain Research Bulletin. **37** 369-375.

Smith, P. (1991) COSHH and the pig industry. Pig Veterinary Journal **28** 103-109.

Stombaugh, D.P., Teague, H.S., Roller, W.L. (1969) Effects of atmospheric ammonia on the pig. Journal of Animal Science. **28** 844-847.

Taiganides, E.P. (1987) Animal waste management and waste water treatment. In: Strauch (Ed.) *World Animal Science.* **B6** Animal Production & Animal Health. Elsevier, Amsterdam.

Taiganides, E.P. and White, R.K. (1969) The menace of noxious gases in animal units. Transactions of the ASAE **12: 3** 359-367.

Tynes, V. (1997) Progression Companion Animal Behaviour. **27** 3. 667-691.

Van Putten, G. (1969) British Veterinary Journal. **125** 511-516.

Wathes, C.M. Demmers, T.G.M., Teer, N, White, R.P., Taylor, J.L., Bland, V., Jones, P., Armstrong, D., Gresham, A., Hartung, J., Chennels, D.J. and Done, S. (2004) Production responses of weaner pigs after chronic exposure to airborne dust and ammonia. Animal Science, **78**, 87-98.

Wathes, C.M., Abeyesinghe, S.M. & Frost, A.R. Environmental design and management for livestock in the 21st century: Resolving conflicts by integrated solutions. Proceedings of the Sixth International Symposium on Livestock Environment **VI** 5-13.

Wells, D.A., Moran, P. (1979) Full-scale experiments to determine the influence of building geometry on wind loads under transverse winds. N.I.A.E. Departmental Note. DN/9/951/04024.

Wemelsfelder, F. (1990) The Experimental Animal in Research. Biomedical CRC Press. **1** 243.

Wood, E.N. (1979) Increased incidence of stillbirths in piglets associated with high levels of carbon monoxide poisoning. Veterinary Record. **104.**

INDEX

Abatement notice 38
Abattoir operators 15
Actinobacillus (Ap)12
Active immunity 47
ADAS Farm Buildings Development
  Centre, 107
ADAS Terrington, 35
Ad-lib fed, 49, 51, 65, 122
Ad-lib hopper, 54, 67
Air inlet, 133, 134
Air movement patterns, 131, 138
Air removal, 104
Air speed, 2, 77, 121, 127
Air velocity, 4
Airborne contaminants, 33
Airborne dust, 97
  concentration, 103
Alarm, 61
Alarm system, 144
Alcohols, 85
Aldehydes, 85
Allergens, 98
All-in/all-out, 81, 102, 110
Ambient temperature, 104, 111
America, 12, 145
Amides, 85
Amines, 85, 87
Amino acids, 35, 64
Ammonia, 35, 85, 87, 88, 89, 102
  concentration, 36, 107
  emissions, 36
  gas, 86
  salt, 86
Anaerobic
  decay, 85
  decomposition, 88
  pig slurry, 85
Animal and Fresh Meat (Examination of
  Residues) Regulations, 19

Antibiotic growth promoters, 18
Antibiotic residues, 16
Antibiotics, 2, 16
Antioxidant, 22
Aqueous ammonia, 35
Artificial environment, 52
Assured British Pigs Scheme, 9
Assured Foods Standards Scheme, 9
Atmospheric particles, 95
Australia, 11
Automatic air inlets, 143
Automatic Control of Natural Ventilation
  (ACNV), 152-155
Automatic variable air inlet, 135
Auxillary ventilation system, 122

Baby pigs, 46, 71
Back-up system, 144
Bedding, 2
Bedding materials, 95
Behaviour, 52
  patterns, 60
  needs, 5
Best Available Technique (BAT), 38
Beta-agonists, 16
Boar pens, 58
Body fat, 47
Bone, 48
BPEX (British Pig Executive), 7
Breeder-feeder, 102
Breeding stock, 50
British Pig Industry Support Group, 6
Broiler feed, 29
Broilers, 23
Bronchitis, 34
Butyric acid, 87

Calculations, 109
Canada, 41, 56

Cannibalism, 66
Carbon dioxide, 85, 87, 90, 91, 111, 112
Carbon monoxide, 85, 92, 93, 112
Carboxyhaemoglobin, 83
Carcinogenic, 22
Centre for Rural Building, 74
Centre-pivot flaps, 152
Chewing, 66
Climate change levy, 36, 154
Coanda effect, 136
Codes of Practice, 22, 25
Cold conditions, 78
Colostrum, 46, 47
Comfort zone, 5, 71
Compassion in World Farming, 8
Concrete floor surfaces, 63
Condemnations, 54
Condensation, 78
Confrontations, 58
Consumer
  attitudes, 2, 4
  demands, 2
Convection, 78
'Cool cell' air inlet, 144
COSHH – Control of Substances Hazardous to Health 27, 33
Courting song, 60
Creep feed, 71, 73
Creep heaters, 79
Crude protein, 35
Curtain-sided building, 151, 152
Cyclone pump, 106

Danes, 6, 18
Danish, 18, 23, 24
Danish Slaughter Houses Association, 16
Danish Veterinary Services, 16
Danske Slagterier, 16, 17. 23
'Dead' areas, 151
'De-animalised meat', 22
Dehydration, 54
Denmark, 6, 16, 18, 24, 120, 122
Department for Environment, Food & Rural Affairs (Defra), 3, 27
Depopulation, 102
De-population, 26
Deposition rates of lean and fat, 49
Differentiated product, 11
Digestible energy, 64, 73

Dissipation of heat, 75
DNA profiling, 22
Dominant pigs, 58, 59, 65
Downdraught jet, 136
Draught-free conditions, 2
Dried blood, 66
Dry sow accommodation, 10
Dry sows, 15
Dunging area, 54
Dust, 4, 82, 95, 102
  concentration, 99, 101
  control, 100
  protection, 83
  removal, 95
Dust-extracted chopped straw, 103
Dust-extracted soft-wood shavings, 103
Dutch Meat Board, 56

E.C. Nitrate Directive, 35
Eating quality, 43
Eaves inlet, 137,138,139
Eggs, 23
Elanco, 13
Electrical
  equipment, 145
  safety, 71
Electrical Safety at Works Regulations, 82
EN 45011, 9
Energy, 49
Enhanced welfare standards, 7
Environment Agency, 38
Environment enrichment, 42
Environment Protection Act, 38
Environmental enhancement, 3
Environmental Health Officer (EHO), 38
Environmental Impact Assessment (EIA), 41
Enzootic pneumonia, 12, 34
Epidemiology, 29
Ethical concerns, 4
EU, 8, 12, 15, 17, 22, 60
EU Council Directive 96/23/EC, 15
Europe, 11
European Commission, 17
Evaporative cooling, 78
Excessive cold, 71
Excessive heat, 71
Excretory behaviour, 53
Extract systems, 142
Extraction equipment, 33

Eye irritation, 93

Face masks, 33
Faecal contamination, 25
Fail-safe device, 143
Fan capacity, 140
Fan size, 140, 141.
Farm assurance, 21
Farm assurance scheme, 9
Farrowing, 15
  pens, 69
  sows, 112
Fat, 48, 50
Feed conversion efficiency, 43
Feed
  hopper, 54
  intake, 76
  particles, 95
  restriction, 60
Fibre, 66
Finisher, 15, 112
Finishers, 119
Finishing pigs, 56, 120, 122
Five freedoms, 7
Flap size, 154
Flavouring compounds, 65
Floor
  surface, 56, 66, 74
  type, 76
  types, 75
Fogging, 106
Food Act, 8
Food chain, 3
"Food Chain Partners", 8
Food
  safety, 9
  scare, 11
Food Standards Agency, 25, 27
Foraging, 52
Freedom Foods, 7
Free-range, 59
Fully slatted, 122
Fully slatted pens, 59

Gables, 149
Gases, 95
Geneticists, genetic selection, 5
Gilt pregnancy, 51
Glucose, 65

Grazing and rooting, 52
Great Britain, 16
Group feeding, 5
Group
  housing, 5
  size, 79
Growers, 119
Growing pigs, 56,118
Growth, 85
Growth promoters, 11
Growth rate, 43

Health & Safety at Work Regulations, 30
Health and Safety at Work Act, 2
Heat loss, 75, 109, 111
Heat output, 74, 111
Heat stress, 53
High air speeds, 47
High humidity, 78
High speed jet, 136,137
Homeotherms, 71
Hot weather, 120,137,144
House-keeping score, 11
Huddling, 63
Human medicine, 12
Hydrogen, 87
Hydrogen sulphide, 85, 91, 92, 93
Hypersensitivity, 22, 34
Hyperthermia, 54

IgE and IgG antibodies, 33
Immunoglobulin, 46
Immunological response, 33
Impact damage, 82
Impeller diameter, 141
Independent auditing, 11
Index of Protection (I.P.) Rating, 82
Indoles, 87
In-feed antibiotics, 12
Inlet area, 127
Inlets and outlets, 130, 131 143 144 146
Inspirable dust, 96, 105
Inspirable particles, 96
Insulated concrete, 75, 76, 78
Insulated walls, 79
Insulation, 2
Integrated Pollution Prevention & Control
  (IPPC), 36, 37, 46, 145, 154
Intensive indoor environment, 5

164 Pig Environment Problems

IPPC, 38, 154
Isothermal, 140

Kennel (s), 79, 102, 116, 147 148
Knock-out facility, 150

Lactation, 50, 51, 85
Lactose, 65
Lean, 48
Lifetime productivity, 50
Light, 71
Light intensity, 81
Linear actuator, 152
Lipid, 45
Lipid protein ratio, 47
Little Red Tractor, 9
Local Planning Authority (LPA), 40
Low air temperatures, 47
Low humidity, 78
Low protein diets, 35, 36
Low ranking pigs, 65
Lower critical temperature (LCT), 63, 71, 72, 73, 74, 77, 79, 80, 114
Lux, 81
Lying area, 54, 136,147
Lysine (see amino acids)

Maintenance, 48
Major retailers, 6
Manipulation, 59
Mark of distinction, 9
Marker Assisted Selection (MAS), 5
Master-timer, 154
Maximum Exposure Limit (MEL), 96
Maximum liveweight, 111
Maximum Recommended Limit (MRL), 16
Maximum ventilation rate, 109, 110,111, 119
Meat hygiene service, 15
Medical tests, 34
Medicated Feeding Stuff Prescription (MFS), 21
Medication, blitz medication, medication strategy, 15
Meteorological data, 154
Methane, 87
Microbial protein, 87
Micro-organisms, 23, 95, 105
Milk production, 81

Minimum ventilation rate, 109, 110, 111, 112, 113, 120, 121
MLC – Meat & Livestock Commission, 9, 27, 29
Moisture resistant, 82
Molecular genetics, 5
MTC stable system, 146
Mycoplasma
  bacteria, 97
  hypopneumoniae, 12

Nasal disc, 43
National Pig Association, 25
National Research Council, 17
Natural ventilation, 151
Naturally ventilated systems, 146, 150
Nausea, 93
Negative pressure, 127, 135
Neonatal pigs, 46
Nest
  building, 69
  occupation, 69
Nipple drinkers, 53, 56
Nitrate pollution, 35
Nitrate Vulnerable Zones (NVZ), 35
Non-organic sources, 22
Non-statutory surveillance, 16
North America, 11
Noxious gases, 4, 85

Odour units, 39
Odours, 38, 93
Oestrus, 68, 81
Optical particle counter, 106
Organic acid, 29, 85
Organic
  production, 22
  waste, 88
Overstocking, 53
Oxygen, 87

P_2 backfat,
Packaging and labelling, 4
Partially slatted, 122
Particle size, 97
Parturition, 69
Passively operated air inlets, 134, 135
Pathogens, 98

Permitted development, 40
PH, 85
Pheromones, 58
Pig
  debris, 95
  farm workers, 33, 34
  slurry odours, 39
Pig Veterinary Society, 25
Pig Welfare Code, 3, 57, 81, 117
Planning Policy Guidance Note PPG7, 40
Pollution of water courses, 3
Polythene duct, 122
Pork, 23
Portal-frame building, 149
Pregnancy, 50, 51
Pregnant sows, 112
Premium prices, premium payments, 6
Pressure changes, 126
Pressure differences, 125, 126, 127, 128, 129, 134
Pressurised systems, 142
Primary dust, 95
Probability and severity rating, 31
Protein, 66
Provision of Antibiotics for Human Treatment Act, 12
Puberty, 51
Public Health Laboratory Service (PHLS), 24, 25
Punched metal, 76

Quality assurance, 3
Quality Bacon Standard, 7, 9
Quality Standard, 10

Radiant effects, 78
Raised floor, 56
Ramp, 56
Rate of air removal, 128
Ration boredom, 66
Rectal prolapses, 54
Relative humidity, 78, 104
Respirable dust, 96, 105
Respiratory diseases, 34, 97
Respiratory pig diseases, 12
Responsible use of Medicines in Agriculture Alliance (RUMA), 20
Restrict fed, 49
Rhinitis, 34

Ridge, 149
Ridge inlet ventilation system, 138, 139
Risk assessment, 30
Risk, risks, risk control, 17
Roller mechanism, 150
Rooting, 59
RSPCA (Royal Society for the Protection of Cruelty to Animals), 7
RSPCA Welfare Standards for Pigs, 7

Safe working practices, 32
Salmonella, 23, 29
Salmonella enteritidis, 25, 29
Salmonella safe, 29
Salmonella typhimurim, 25
Salmonellosis, 13. 23
Scandinavia, 17, 22
Scottish Centre for Infection and Animal Health, 24
Secondary dust, 95, 101
Sedimentation, 95
Selection, 51
Selective breeding, 43
Sensible heat loss, 113, 114
Service, 51
Settling velocity, 99
Sexual behaviour, 68
Sexuality, 69
Shade, 63
Shy pigs, 58
Silsoe Research Institute, 137, 145
Skatole, 87
Slat level, 87
Slaughter pigs, 46
Slurry channel, 103
Social
  groups, 58
  hierarchy, 58
  integration, 69
Sodium ascorbate, 22
Sodium nitrite, 22
Solid waste, 103
Sow feeding, 51
Sow milk, 14, 46
Sow stalls, 5
Space-boarding, 147
Split level pen, 56
Spot-checks, 9
Sprinkler cooling system, 122, 123, 124

Stable groups, 59
Stack-effect, 128-130
Standing Committee of the European Convention for the Protection of Animals Kept for Farming Purposes, 58
State veterinary service, 15
States, 12
Statutory Surveillance Programme, 16
Stereotypical behaviour, 65
Stocking density, 53, 66, 116
Stockmanship, 52
Stotfold Pig Development Unit, 101
Straw, 60, 75
Straw bed, 75
Stress, 66
Sucrose, 65
Suffocation, 54
Sulphides, 85
Sulphonamides, 16
Supermarket shopping trolleys, 6
Supermarkets, 6, 11
Supplementary heat, 110
Surface area, 79
Sweden, 16

Tail biting, 67
Tail docking, 67
Temperature, 60, 63, 71,75,79,85,128, 129
Temperature lift, 4
Thermal
  barrier, 76
  comfort, 71
  environment, 56
Thermostat, 153
Toxins, 23
Traceability and due diligence, 9
Tunnel ventilation, 141

UK, 3, 5, 6, 7, 8, 16, 25,36, 37 38, 40, 145
Upper Critical Temperature (UCT), 71, 72, 76, 111, 114, 118
Urea, 35, 87
Urease, 87
Uric acid, 86
USA, 8, 23, 142,146
USA's National Research Council, 17

Value-added products, 4
Vapour pressure, 85
Varkensbesluit, 56, 59
Ventilated ridge, 99,104,148, 149,
Veterinary Laboratories Agency, 29
Veterinary Medicines Directorate (VMD), 15, 21
Veterinary residues, 16
Vices, 54, 66, 120
Visual barriers, 58
Vocalisation, 60
Volume throughput, 140

Wallowing, 120
Warm-blooded, 71
Water
  bowls, 53
  courses, 35
  guage, 129
Weaned sows, 112
Weaners, 15, 59, 112
Weaning, 51
Welfare needs, 4
Welfare of Farmed Animals (England) (Amendment) Regulations, 81
Welfare of Farmed Animals (England) Regulations, 80, 143
Wet fed, 122
Wet feed, 29
Wet scrubber, 107
Wheat Treet, 29
'White-glove clean', 82
Wholesome, 3
Wholesomeness, 2, 16, 22
Wild boar, 43
Wild pigs, 60
Wind, 126, 127
Wind pressure, 126
Withdrawal period, 19
World Health Organisation (WHO), 20
WTO Round, 8
WTO (World Trade Organisation), 8

Zoonosis Action Plan (ZAP), 27, 29